THE WORLD IN CONFLICT

War Annual 6

THE WORLD IN CONFLICT

Brassey's titles of related interest

JOHN LAFFIN
Brassey's Battles: 3,500 Years of Conflict, Campaigns and Wars from A–Z
JOHN LAFFIN
War Annual 1
War Annual 2
War Annual 3
War Annual 4
War Annual 5

Also by John Laffin

Military

Middle East Journey
Return to Glory
One Man's War
The Walking Wounded
Digger (The Story of the Australian Soldier)
Scotland the Brave (The Story of the Scottish Soldier)
Jackboot (The Story of the German Soldier)
Tommy Atkins (The Story of the English Soldier)
Jack Tar (The Story of the English Seaman)
Swifter than Eagles (Biography of Marshal of the R.A.F. Sir John Salmond)
The Face of War
British Campaign Medals
Codes and Ciphers
Boys in Battle
Women in Battle
Anzacs at War
Links of Leadership (Thirty Centuries of Command)
Surgeons in the Field
Americans in Battle
Letters from the Front 1914–18
The French Foreign Legion
Damn the Dardanelles! (The Agony of Gallipoli)
The Australian Army at War 1899–1974
The Israeli Army in the Middle East Wars 1948–1973
The Arab Armies in the Middle East Wars 1948–1973
Fight for the Falklands!
On the Western Front: Soldiers' Stories 1914–18
The Man the Nazis Couldn't Catch
The War of Desperation: Lebanon 1982–85
Battlefield Archaeology
The Western Front 1916–17: The Price of Honour ⎫
The Western Front 1917–18: The Cost of Victory ⎬ Australians at War
Greece, Crete & Syria 1941 ⎭
Secret and Special
Holy War: Islam Fights
World War 1 in Postcards
Soldiers of Scotland (with John Baynes)
British Butchers and Bunglers of World War 1
The Western Front Illustrated
Guide to Australian Battlefields of the Western Front 1916-1918
Digging Up the Diggers' War
Panorama of the Western Front
Western Front Companion

General

The Hunger to Come (Food and Population Crises)
New Geography 1966–67
New Geography 1968–69
New Geography 1970–71
Anatomy of Captivity (Political Prisoners)
Devil's Goad
Fedayeen (The Arab–Israeli Dilemma)
The Arab Mind
The Israeli Mind
The Dagger of Islam
The PLO Connections
The Arabs as Master Slavers
Know the Middle East
Fontana Dictionary of Africa since 1960 (with John Grace)

And other titles

THE WORLD IN CONFLICT

War Annual 6

Contemporary warfare described
and analysed

JOHN LAFFIN

BRASSEY'S
LONDON • WASHINGTON

COPYRIGHT © 1994 John Laffin

All rights Reserved. No part of this publication may be reproduced, stored in a retrieval system or transmitted in any form or by any means: electronic, electrostatic, magnetic tape, mechanical, photocopying, recording or otherwise, without permission in writing from the publishers.

First English edition 1994

UK editorial offices: Brassey's, 33 John Street, London WC1N 2AT
UK orders: Marston Book Services, PO Box 87, Oxford OX2 ODT

USA orders: Order Department, Macmillan Publishing, 201 West 103rd Street, Indianapolis IN 46290, USA

Distributed in North America to booksellers and wholesalers by the Macmillan Publishing Company, NY 10022

John Laffin has asserted his moral right to be identified as author of this work.

Library of Congress Cataloguing in Publication Data
available

British Library Cataloguing in Publication Data
A catalogue record for this book is available from the British Library

Hardcover 0 08 0413307

Printed and bound in Great Britain by
Butler & Tanner Ltd, Frome and London

For my wife, Hazelle, my assistant in the difficult and sometimes dangerous task of gathering information for War Annual. Since she also prepares the finished typescript, every entry passes through her hands more than once and, in effect, she performs the duty of 'quality control'. She insists on verification of information received, though this is not always possible. Many an author, in his Acknowledgements, writes that without the help of somebody or other 'this book could never have been written'. Never more truly has this been so than in the case of Hazelle Laffin and War Annual.

CONTENTS

List of Maps — x

Introduction — xiii

Afghanistan — 1
   A War of Hatred and Fear

Guerrilla Civil War in Angola — 13
   Peace – Perhaps

Bangladesh War of Genocide — 21
   Time Runs Out for the Resistance

Burma (Myanmar) Guerrilla War — 25
   The Karens' Fight for Freedom

Cambodia's On-Off-On War — 35
   Enter the United Nations

The Central American Arena — 44

El Salvador Civil War — 47
   Towards a Peaceful End

Guerrilla War in Guatemala — 52
   Professional Rebels

Nicaragua – The Contra War — 56
   Conflict in its Death Throes

Colombia's Civil War — 59
   'The War that Will Not End'

East Timor Resistance War	67
Indonesia's Unrelenting Oppression	
Ethiopia – Eritrea – Tigre – Somalia	73
India – Pakistan War	89
Indian Army's Terror Campaign	
Iraq War (or Second Gulf War)	96
The Profit and Loss Account	
Israel and the Palestinians	110
The Intifada (Arabic for 'shaking free')	
Kurdish War of Independence	118
New Phase	
Looking for Peace in Lebanon	133
The Syrian Coup	
Morocco – Polisario War	144
Strains on the Referendum	
Mozambique Guerrilla War	149
Moving Violently Towards Peace	
Northern Ireland Terrorist War	156
The Libyan Connection	
Papua New Guinea	165
Island Rebels	
Peru's 'Shining Path' War	168
A Country in Chaos	

Philippines 'People's War'	174
A Nest of Conflicts	
South Africa: Zulu – ANC War	179
The 'Black-on-Black' Conflict	
Sri Lanka Civil War	186
'Tiger, Tiger, Burning Bright'	
Sudan Civil War	195
Brigadier Bashir's 'Fantasy World'	
Sumatra: Separatist War in Aceh	201
Civil War in former Yugoslavia	203
Ancient Hatreds, Fresh Grief	
War Trends – Racing Towards Conflict	220

Maps

The former Soviet Union's Republics
Autonomous Republics of the CIS
U.S Troops Around the World
The Afghanistan Resistance
Angola: Areas Where Fighting Has Died Down
Bangladesh Guerrilla War
Burma (Myanmar) Guerrilla War
Cambodia: Khmer Rouge Activity Continues Despite Peace Efforts
Central America
El Salvador Civil War
Guerrilla War in Guatemala
Nicaragua: Sandinistas *versus* Contras
Colombia Guerrilla and Narcotics War
East Timor Resistance War
Ethiopian Liberation Wars at an End
Ethiopian Tribal Areas
Battle Area (Ethiopia)
Rebel Territories (Ethiopia)
Somalia
India — Pakistan Confrontation
Conflict in India
The Iraq War
   Options for UN Attack in February 1991
   Iraqi Deployments
The Iraq War (General Map)
West Bank and Gaza: Areas of Intifada
Kurdish War of Independence
Saddam's War Against the Kurds: Allied Protection
'Happy Valley' Safe Haven
Lebanese Regions of Conflict as Syria Extends Control
The Jezzine Salient
Morocco—Polisario War
Mozambique Civil War
Northern Ireland
Peru: Sendero Luminoso (Shining Path) War
Philippines Ethnic Areas
Philippines Guerrilla War
Zulu — ANC Battlefields
Sri Lanka: Ethnic and War Areas
Elephant Pass
Sudan Civil War

Yugoslav Civil War (August 1991)
Yugoslavia Divided
Croatia: Ethnic Groups
Bosnia
NATO Airstrikes: The Final Countdown
The Battle for Gorazde
Syria's North Korean Missiles
Spratly Islands and the Six Nations which Claim Them
Regions with Uncleared Land Mines

Introduction

A WORLD WITH ONE SUPERPOWER

The war fought against Saddam Hussein's Iraq — popularly called the Gulf War but which should be styled the Iraq War — was as much a watershed in military history as was the breaking down of the Berlin Wall in political history. Military and political history are not really divisible but the two events were significantly different in that in Iraq and Kuwait we saw much loss of life and destruction of a country, while with the destruction of the Wall an ideology was discredited and tremendous social and political change occurred.

In, *War Annual 5*, my Introduction had the rhetorical sub-heading 'Who said War is less likely?' and I argued that with only one superpower in existence there would be more war than when there were two superpowers. In that strategic balance each of them was uneasily aware that the other had an awesome power of military veto over its actions. For the entire period of the Cold War, the United States and the Soviet Union were forced to make plans with each other's probable reaction in mind. No significant movement by sea, land or air could be contemplated without the question being asked, in the Pentagon or the Kremlin, 'How will the Soviets (or the Americans) react to this? Will they go along with it? Will they regard it as hostile?' There was always apprehension, often tension, sometimes acute anxiety.

When Iraq invaded Kuwait and precipitated an international crisis the US was under no constraint from anxieties about what the Soviet leadership might do if the US went to war against Iraq. The two powers 'consulted' at presidential and ministerial level but at no time did the Americans feel the need to seek Soviet approval and consent for their actions. If that was given, well and good but the Americans were merely *informing* the Soviet leadership for the sake of good relations.

The much-publicised summits between the American and Soviet leaders were intended to influence world opinion in general, Middle East and Islamic opinion in particular and Saddam Hussein most of all. The implicit message was: 'In this grave crisis, with a dangerous bully threatening the peace of the world, the two superpowers stand together against him.' This union was impressive. The Soviet Union had, after all, been the friend of Iraq and had been Saddam's principal supplier of arms, intelligence and training. Now it was not only repudiating its obligations to Iraq but joining an old enemy in virtually common purpose against Iraq.

Some world leaders may have believed that President Gorbachev and his colleagues were showing statesmanlike qualities in acting together with the Americans for the common good of humanity. Perhaps they were, but not from any altruistic motive. They had a simple option: They could clamber onto the stage in the hope of being seen in the spotlight that was focussed on the Americans — and which was spreading some glow on the US's allies — or they could sulk in impotent isolation and allow matters to take their course. It was obviously better to pretend to be a leading player with a major voice-part than a walk-on bit player carrying a fake spear.

The US Administration was remarkably generous in allowing the Soviets to enhance their reputation in this way. President Gorbachev, as leader of a former superpower that was fast fragmenting, understood his limitations and he was grateful to President Bush for the reprieve he had been granted. For a little while longer, he could pose as a Grand Leader.

The Americans understood the significance of their change of role to single superpower very well. When the Saddam Crisis occurred, nobody in the Bush Administration had any doubt that only the US could throw the Iraqi dictator's ambitions into reverse. For the sake of world opinion, and to neutralise the inevitable anger of the Arab and Islamic nations at having one of their number attacked, it was necessary for the Americans to seek United Nations' approval for sanctions and for military action. It was also necessary to have Western allies with military muscle. It was equally necessary for the US to have partners in the enterprise in order to augment the moral authority for the war against Iraq. In fact, the moral justification was hardly in doubt, whatever other reasons for war might have been taken into account.

The US leaders handled the preparations for war so efficiently that they induced some Arab states to join the coalition against Iraq. Some needed no persuasion at all. Saudi Arabia's leaders felt so directly threatened that their participation in the war against Iraq could be taken for granted. Syria was a different proposition. The Americans hated Syria and its brutal ruler, Hafez al-Assad, a terrorist plotter and paymaster, but his presence in the coalition was considered necessary in order to show that even enemies could unite against an aggressor as vile as Saddam Hussein.

Assad was told that if he were to send a token military force to Saudi Arabia to join the Western powers in throwing Saddam out of Kuwait, the US would be more co-operative in the future. For instance, the Americans would put pressure on Israel to take part in Middle East peace talks with Syria. Assad sent to Saudi Arabia an armoured unit of 4,000 men who did nothing more than garrison a section of the line — actually well removed from the front — where they were never in danger.

Before the US went to war against Saddam, the Bush Administration had fulfilled every precondition necessary for a war-making power to be confident of success. These were among its most important prelude-to-war achievements:

- It could count on certain powerful and influential Western allies, particularly Britain and France.
- It had the formal approval of the United Nations General Assembly and the Security Council.
- It had been invited by Saudi Arabia to protect that nation's territory, which meant the US army had 'permission' to be in the country in strength. It had similarly been 'invited' by Kuwait to recover its sovereignty, stolen by Saddam Hussein.

- It had eliminated the possibility of tensions arising in its relationship with the Soviet leadership.
- It had neutralised Saddam's natural allies — the other Arab states and the Islamic world as a whole.
- Finally, and very importantly, Bush and his team had secured popular consensus at home to go to war for a cause that did not directly affect the United States: neither its citizens nor the soil were under threat.[1]

All this was a masterly performance which proved that American diplomacy had become as sophisticated as that of the much older powers.

It is diplomatically correct to call the military action taken to drive Saddam Hussein and his army from Kuwait 'the UN war against Iraq'. In all essentials it was the United States' war and its operations and strategic direction were wholly under American direction and nearly always under American command.

This is not to deny that from time to time strains and difficulties occurred but on the whole few countries throughout history have ever embarked on a war with such complete political, psychological, material and strategic preparation and therefore with a justifiable expectation of winning the conflict. Bush and his colleagues and allies had only to sustain the domestic and foreign consensus which alone provided the green light for maintaining the war.[2]

American military planners think big — hence the tremendous build-up of half-a-million men in the Saudi deserts, together with the vast quantities of arms, ammunition and supplies needed for the type of war being planned. Iraq's military installations as well as the infrastructure that supported them — such as roads, bridges and communications links of all types — would be so heavily pounded that the Iraqi forces, large and powerful though they were, would be isolated on the battle fronts. Reinforcements and fresh supplies could reach them only with great difficulty and their morale would collapse.

The American plans called for the enemy to be crushed, not merely beaten. The thinking behind this intention was that Saddam would be forced to pull his army of occupation out of Kuwait or witness its destruction. In the end, both happened — much of the army was destroyed while pulling out.

The stated objective of the war — to expel the Iraqi forces from Kuwait — was achieved. The war itself was prosecuted with a speed, vigour and force never before equalled — though there have been smaller campaigns *within* wars that were incisive and decisive. The allied onslaught resulted in massive casualties to the aggressor Iraqis, many of them civilians, and very few casualties to the defenders and liberators. In all these particulars the Iraqi War will set an example and be a model for the 21st century.

The doctrine of 'minimum force' lost its relevance during the Iraq War. The concept of overwhelming force to accomplish a stated objective quickly, though not new in itself, gained new credibility, at least in military minds. During 1991 it was the theme of lectures in armed forces colleges around the world.

The great paradox is that while the US and its allies won the war they did not defeat Iraq. Its detested leader Saddam Hussein remained in power. Various reasons have been given to explain why President Bush ordered the operations to conclude when they did, after only eight days ground fighting. These are some of them:

- For Western Christian troops to enter the great Islamic city would have angered the Muslim world.
- The UN resolutions authorising the war were limited to removal of Iraq's troops from Kuwait; when this was achieved there was no mandate to continue the war.
- To interfere in the internal affairs of any country was to breach the UN Charter.

All these explanations are valid but another reason for the end of the US and Allied assault was that President Bush, having achieved so much with the loss of so few American servicemen and servicewomen feared that heavy casualties might result from further fighting. He said as much in a national televised broadcast on 1 March 1991. 'I did not want to lose any more of our boys.'

It was understandable that, having lived through the Vietnam War and its aftermath, he found difficulty in facing the possibility of electoral defeat at the hands of an American public suffering from an 'Iraq trauma.'

In his moment of triumph, the President announced, 'I have kicked the Vietnam syndrome.' His statement showed to what a great extent the syndrome had inhibited him but many observers wondered if he had indeed kicked it. It seems possible that even as he expressed great national pride in American achievements in the Gulf he was still in its grip. This seemed to be so from the next developments in Iraq.

The Shia people of southern Iraq and the Kurds of the north rebelled against Saddam Hussein and the US–UN forces did not go to their aid. President Bush was aware that 75 per cent of the American public were opposed to further military action in the region. Also China, the Soviet Union and many Third World countries such as Indonesia opposed action in Iraq in support of minorities because they have their own serious problems with minorities and did not want to create a precedent for US or UN intervention on their behalf.

It is my belief that had the war continued for only a few more days — as General Schwarzkopf wanted — then Saddam would have fallen to his own people, perhaps even in a coup organized by some of his lieutenants. Not all military analysts would agree with me on the certainty of Saddam's fall at this time and it will forever remain a cause of tantalising speculation.

Within the Arab world and the wider Muslim world, Saddam is perceived to have *won* the war. This view is based on the simple logic of the tradition in Arab–Islamic history that a defeated leader is either dead or is a prisoner in a cell or has fled to a friendly country for refuge. Clearly, none of these fates had befallen Saddam. He remained in charge of a government and leader of the armed forces. In addition, the 'invaders' had left Iraq. By the same simple, traditional logic this meant that they had been defeated. Victorious armies do not quickly move out.

Elsewhere, the the war left the US with an awesome military reputation: even its detractors conceded this. Much of the arms and equipment had been used in battle for the first time in Iraq and their sophistication and technological brilliance was breathtaking. In contrast, the Soviet equipment used by the Iraqi forces was shown to be inferior.

THE WORLD IN CONFLICT

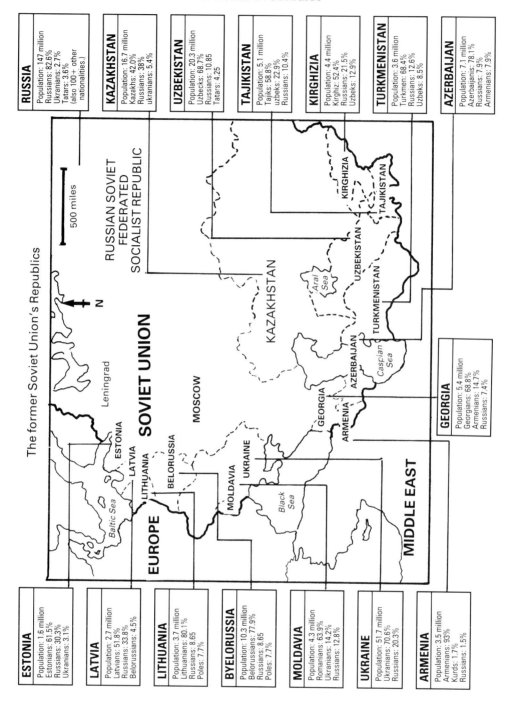

The former Soviet Union's Republics

RUSSIA
Population: 147 million
Russians: 82.6%
Ukranians: 2.7%
Tatars: 3.6%
(also 100+ other nationalities.)

KAZAKHSTAN
Population: 16.7 million
Kazakhs: 42.0%
Russians: 38%
ukranians: 5.4%

UZBEKISTAN
Population: 20.3 million
Uzbecks: 68.7%
Russians: 10.85
Tatars: 4.25

TAJIKISTAN
Population: 5.1 million
Tajiks: 58.8%
uzbeks: 22.9%
Russians: 10.4%

KIRGHIZIA
Population: 4.4 million
Kirghiz: 52.4%
Russians: 21.5%
Uzbeks: 12.9%

TURKMENISTAN
Population: 3.6 million
Turkmen: 68.4%
Russians: 12.6%
Uzbeks: 8.5%

AZERBAIJAN
Population: 7.1 million
Azerbaijanis: 78.1%
Russians: 7.9%
Armenians: 7.9%

GEORGIA
Population: 5.4 million
Georgians: 68.8%
Armenians: 14.7%
Russians: 7.4%

ESTONIA
Population: 1.6 million
Estonians: 61.5%
Russians: 30.3%
Ukranians: 3.1%

LATVIA
Population: 2.7 million
Latvians: 51.8%
Russians: 33.8%
Belorussians: 4.5%

LITHUANIA
Population: 3.7 million
Lithuanians: 80.1%
Russians: 8.65
Poles: 7.7%

BYELORUSSIA
Population: 10.3 million
Belorussians: 77.9%
Russians: 8.65
Poles: 7.7%

MOLDAVIA
Population: 4.3 million
Romanians: 63.9%
Ukranians: 14.2%
Russians: 12.8%

UKRAINE
Population: 51.7 million
Ukranians: 70.6%
Russians: 20.3%

ARMENIA
Population: 3.5 million
Armenians: 93%
Kurds: 1.7%
Russians: 1.5%

xvi

Autonomous Republics of the Commonwealth of Independent States (CIS)

Autonomous Republics in Russia	
1 Kabardino-Balkar	5 Kalmyk
2 North Ossetia	6 Mordovia
3 Chechen-Ingush	7 Chuvash
4 Daghestan	8 Mari

—··— Russian Republic (Western boundary)

While the Soviet Union continued to disintegrate to the point where the 'Union' ceased to exist, the prestige of the United States grew. By late 1991 the American leadership no longer saw the necessity of 'consulting' with President Gorbachev, while he clung on as leader of a 'phantom empire' for as long as he could. They did not even indulge in the formality of 'informing' him of what they proposed to do in the Middle East, Asia, South-east Asia and elsewhere. This courtesy they transferred to the presidents of Russia, Ukraine and Kazakhstan, the dominant nations emerging from the ashes of the Soviet Union. All of them were eager to be friends of the Americans because all had learnt two basic lessons from the Iraq War: first, that the United States can, after all, win a great war quickly and ruthlessly; and second, that no nation and no combination of nations can stop them. It makes good sense to be friendly with the Americans.

Students of war need to study the Arab reaction to the war. It is not enough to look at it from a Western perspective. By mid-1992, several books had been published in Arabic but the most important and most cogent in English is that by Mohamed Heikal, the Arab world's leading commentator and journalist.

His book is entitled *Illusions of Triumph* and the title aptly summarises Heikal's opinion that the Western victory was illusory because it left the Arab world as a whole humiliated and therefore vulnerable to further unrest and instability.[3]

Heikal explains that every Middle East war has had a character of its own. In the case of the 1956 conspiracy by Israel, Britain and France to seize the Suez Canal, that character was outrage; in the 1967 war (the Six-Day War) it was humiliation; in 1973, when Egypt and Syria showed that Israel was not invincible, it was elation and pride. However, in what Heikal calls 'the Kuwait conflict' there was no pride, much anger and 'a wrenching of the soul' for the Arab world. Every individual Arab felt an inner division.[4]

Ever well-informed, Heikal reports an incident among a group of Egyptian and Syrian soldiers stationed among the Coalition forces in Saudi Arabia when they heard the news, on 18 January 1991, that Iraq had launched its first Scud missiles against Israel. They were supposed to be enemies of Iraq but they shouted '*Allah Akbar!*' (God is great) in support of Iraq. Seven Egyptians and several Syrians were disciplined by their respective commanding officers but their spontaneous reaction to the news of the attack on Israel showed where their true sympathies lay.

'Nothing is more urgent than the need to contain and defuse the anger of the Arab world, yet nothing could be harder to achieve,' Heikal warns. 'The events of August 1990 to March 1991 left wounds which will take years to heal.'

He believes that the Arabs will inevitably take refuge in Islamic fundamentalism because every other ideology has failed and a return to religious roots is a temporary refuge from current humiliation. 'The only solace and defence is religion, even if it provides no real solution to current problems.'

Russia's New Empire: The West's Humiliation

With the downfall of President Gorbachev and the rise of Boris Yeltsin in Russia, together with the defeat of President Bush and the arrival in the White

House of Bill Clinton, a new era can be said to have begun. The tough, obdurate Yeltsin and the hesitant, weaker Clinton faced enormous problems in their respective spheres of influence and in the world at large.

Neither of the leaders created much confidence. Yeltsin sat uneasily on his throne and had to survive a violent insurrection in the Russian White House itself when his foes tried to unseat him. President Clinton was safe from such an ugly experience but his inadequacy in international affairs worried his Western partners. He was strong enough to order an attack on Baghdad in an effort to force Saddam Hussein to obey UN resolutions but he acted without the conviction that George Bush had shown. He waxed hot and cold over US, UN and NATO intervention in the former Yugoslavia, leaving his principal *confrères* in London, Paris, Madrid and Rome confused about his will and his wisdom. His experience in Somalia was unhappy; his military suffered casualties and failed to capture the warlord General Aidid, a main objective of the temporary occupation.

By 1994 there were signs that Russia was attempting to regain its world power status. An assertive military doctrine was in place as was, unfortunately, Vladimir Zhirinovsky, a vocal Yeltsin opponent with wild imperialist ambitions. Russia should be closely watched, said US Defence Secretary William J. Perry. So, incidentally, should William J. Perry, who virtually invited the Serbs to attack Gorazde when he announced that UN/US airpower would not deter the Serbs.

In 1994 Russia had about a quarter of the number of people in uniform that the Soviet Union had had in 1991 - 1.5 million, compared to 1.7 million for the US, 2.5 million for NATO, excluding the US, and 3 million for China. Russia is not a serious threat to the West. The entire military is in a state of organisational chaos, as are the industries which serve the forces. In the winter of 1993-94 the army and navy reported deaths from starvation and exposure.

However, Russia is still able to build up a post-Soviet empire by roping in its neighbours. Russia's leaders and its bureaucracy possessively refer to the 14 republics as 'the near abroad' and in the Baltic countries, Belarus, Ukraine and across Central Asia, Russia is busily restoring its influence. The strategy is to apply pressure along the ethnic fault lines and play rival factions against one another. Also, there is a tactic of 'protecting ethnic Russians', a process that legitimises military activity. In December 1992 Russian planes from Uzbekistan helped bring down the government of Tajikistan, composed of Islamic and democratic groups, and installed pro-Communist rulers in its place. Elsewhere, Russia hits at the republics' economies as a way of saying 'Russia rules, okay?'. It did this early in March 1994 when it cut off the gas supplies in Belarus, Moldova and Ukraine.

With the help of arms supplied by Russian military commanders, Abkhazia, a part of Georgia demanding independence, overwhelmingly defeated Eduard Shevardnadze's Georgian army, which retreated to Tbilisi. Almost at once Shevardnadze received weapons from Russia, which he used to put down an insurrection led by Zviad Gamsakhurdia. In this way, Russia made it clear that everything in Georgia depends on Russia.

Even Foreign Minister Andrei Kozyrev, regarded in the West as a liberal, turned into a belligerent nationalist in 1994, using rhetoric that reminded

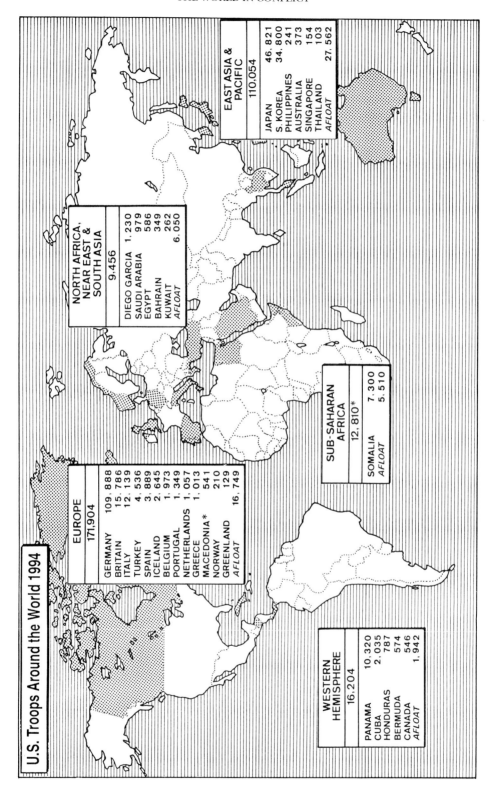

observers of Cold War days. He warned Eastern European countries against joining NATO and yet again threatened the republics if they mistreated ethnic Russians. Russia already has peace-keeping operations in Georgia, Tajikistan and Azerbaijan and, with US approval, sent 400 troops to Bosnia, where they persuaded the Serbs to open the airport in the besieged town of Tuzla.

Whatever the trials, tribulations and triumphs of the US and Russia and their respective leaders, they have much to gain from co-operation. A general outbreak of war in the Balkans would be damaging to the US and to Russia; a war with North Korea over its nuclear armaments would adversely affect everybody, as would any spread of conflict in the Middle East.

From the end of February 1994, Clinton and Yeltsin and the leaders of all other countries had to accept that the world has changed. NATO, the strongest military alliance in history, guardian of the West, fired its first ever shots in anger. Its warplanes shot down four Serbian planes over central Bosnia Herzegovina, thus changing the face of international relations. It was an astonishing event. Throughout its 45 years of existence, NATO had been conditioned to think that its forces would operate on the north German plains, on the southern flank in Turkey, even in the Arctic. The Balkans were hardly considered.

The accepted potential objectives for NATO were always reasonable and logical, but in the Balkans Western policy has always been vague and confused. For instance, in 1992 the West, with much rhetoric, assumed the mantle of Bosnia's protector but then refused to come to its defence. This confused the rest of the world.

Having signalled how far it was prepared to go in shooting down the Serbian planes in February 1994, the West had to be prepared to go further still. Single-mindedly, the West, NATO, the UN and the US, in various combinations, protected food and relief convoys. In April, NATO, with UN approval and the use of American and British warplanes, struck the Serbs again and shot down one of their bombers near Gorazde. Other strafing NATO aircraft tried to hit Serbian tanks, but without effect: the precision bombing of Iraqi targets, during the Gulf War was not repeated. With inadequate support from ground observers - they did not have the correct equipment - the pilots could not find their targets because of heavy clouds. When US Defence Secretary Perry assured the Serb military leaders that they had nothing to fear from the air, the Serbs moved in for the kill at Gorazde.

The West, the UN, the US and NATO will need years to live down their humiliation in Bosnia. Their lack of clear policy and strategy, their inept tactics on the ground and their complete lack of decisiveness - the principal element of leadership - are faults obvious to all the world's ambitious warmongers.

References

1. Catherine M. Kelleher of the Brookings Institute posed the great question for Americans: 'Why us, why now and why there?' She said that this subject of debate would loom larger in American society and would recur in the presidential election of 1992. She was speaking during a seminar at King's College, London, 6 February 1991.
2. Whether Kuwait was worth defending and whether the restoration of its despotic leaders was worth the loss of life among Western soldiers is a matter of debate, largely philosophical, that I cannot address in a book about the war itself.

3. *Illusions of Triumph: An Arab View of the Gulf War*, Mohamed Heikal, (Harper Collins, April 1992). Nobody could be better qualified to speak about the Arabs. From 1957 to 1974 he was editor-in-chief of the Cairo daily newspaper Al-Ahram. A close friend of President Nasser, he was also a confidant of President Sadat until the mid-1970s when they disagreed over Sadat's policies after the 1973 Arab–Israel War. Heikal was among prominent Egyptians imprisoned by Sadat in September 1981. He was released two months later, following Sadat's assassination. He has written many books on Middle East affairs.
4. The comments which follow were made during conversation on 10 March 1992, not in the Heikal book.

Afghanistan

A WAR OF HATRED AND FEAR

Background Summary

This conflict has its roots in the invasion of Afghanistan by the Soviet Union in December 1979. The stated reason was to help the Afghan government 'to maintain control over rebellious elements'. In fact, the Soviet leadership wanted to protect its investment in Afghanistan, one of its satellites. Control of Afghanistan was seen as essential in the traditional Russian policy of pushing towards the Indian Ocean.

When President Hafisullah Amin was killed in the fighting, Babrak Karmal, a Soviet nominee, succeeded him. Karmal could not quell the Resistance, which consisted of Mujahideen, or Islamic holy war warriors, in various groupings. The Soviet leadership sacked Karmal and appointed as president Dr. Muhammad Najibullah, a former head of KHAD, the secret police.

Despite massive help from the US and Pakistan, the divided Resistance could make no progress. As the war destroyed the country's economy, millions of desperate Afghans fled, mostly to Pakistan. Nevertheless, Stinger aircraft missiles from the US and the Blowpipe missile from Britain gave the guerrillas an advantage.

When the Mujahideen groups became more successful in co-ordinating their operations Soviet losses became intolerable and the Moscow leadership decided to withdraw from Afghanistan. The Geneva agreement on 14 April 1988 confirmed that intention. The last units of the Soviet army of occupation left on 15 February 1989 but the Soviet Union continued to support Najibullah with money and weapons.

The Soviet army's casualties were said to be 13,833 and this seems to be more accurate than the 50,000 estimated by Western analysts. At least one million Afghans, mostly civilians, lost their lives and five million more were refugees. For the Soviet Union, the war in Afghanistan was a military defeat, but propaganda at home largely concealed this unpalatable fact. Even in the West there is the feeling that the Soviet army pulled out as part of President Gorbachev's new and enlightened policies rather than because it had suffered a defeat. The fact is that the Soviet army in Afghanistan used totally inappropriate tactics and achieved none of its strategic aims.

Najibullah's War Against the Resistance 1989–90

With the withdrawal of the Soviet army, the war became a direct conflict between the army of the Democratic Republic of Afghanistan (DRA) and the Mujahideen Resistance, which expected a quick victory. Foreign diplomats in

Kabul and in Pakistan predicted that the seven-group Mujahideen Alliance, based in Peshawar, would become more cohesive in their common cause.[1]

They not only failed to do this but they abandoned guerrilla tactics and besieged Jalalabad, Afghanistan's third largest city, on 6 March 1989. They fought for it until July, losing 500 men killed and 1,500 wounded, and conceded defeat. It is interesting to note that the reverse was more the fault of the Pakistani Military advisers than of the Mujahideen. Despite many lessons from war history, they had still failed to understand that artillery barrages and mass assaults could not dislodge a well-entrenched enemy.

Najibullah's prestige rose with the success of the DRA army and during shrewd political manouevres he struck separate deals with various Mujahideen leaders and so divided the Resistance. Najibullah already had a formidable army and it grew in strength because of the vast amounts of equipment handed over by the departing Soviets.

In 1990 he created the Special Guard, a mobile force of 11,000 selected men, and enlarged the paramilitary KHAD to 35,000. These units and his political successes did not protect Najibullah from coup attempts and on 13 March 1990 General Shah Nawaz Tanai, the Minister of Defence, attempted to overthrow him on behalf of a fraction of the People's Democratic Party of Afghanistan (PDPA). MiGs flown by insurgent pilots made 36 sorties against government targets but the rebels were routed by loyalist forces and General Tanai fled to Pakistan. Backed by the Pakistanis, Tanai had been working with Gulbuddin Hekmatyar, leader of *Hezb-i-Islami*.

Najibullah appointed a dynamic new defence minister, General Muhammad Aslam Watanjar. Aged only 43, he had been highly regarded by the Russians and, like Tanai, he is a Pushtun from Paktia province.

Tanai might have failed in his coup but his actions split the Afghanistan military machine. The American CIA and the Pakistan's Inter-Service Intelligence (ISI), working together strove to exploit this advantage by urging the Mujahideen to go on the offensive at once.

They failed to do so and, through tribal feuding, assassinations and general incompetence, they forfeited the good opinion of their backers. Only four states recognized the 'Afghanistan Interim Government' (AIG) — Saudi Arabia, Bahrain, Sudan and Malaysia.

In June 1990, Sibghatullah Mojaddedi, president of the AIG and chief of the Afghan National Liberation Front, accused Gulbuddin Hekmatyar of having ordered the assassination of several Mujahideen notables, as well as two attempts on his own life. During the year more than 30 Mujahideen leaders were murdered in Peshawar alone. The most significant victim was Naseem Akhundzadah, the most powerful chief of southern Afghanistan.

In mid-1990 Washington abandoned hopes of a Mujahideen victory and instead concentrated on a peace plan in conjunction with the Soviet Union. This decision alarmed some Mujahideen field commanders, several of whom tried to co-ordinate their operations against the DRA despite the wishes for separate activity by their own political leaders.

About 40 commanders, mostly Pushtuns, held a meeting in the province of Paktia to discuss armed co-operation. However, the real power lay in Peshawar where the political leaders sabotaged all efforts by their field commanders to

AFGHANISTAN

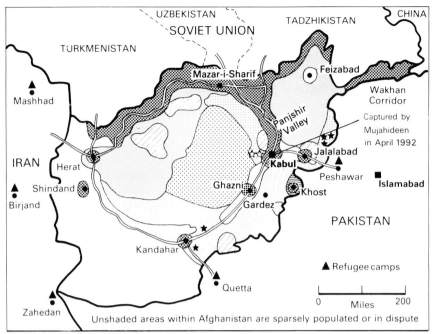

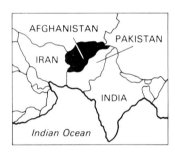

The Afghanistan Resistance

Fundamentalists

☐ Jamiat-i-Islami: Led by Ahmad Shah Massoud. Regarded as the best fighters.

☐ Hezb-i-Islami; led by Younis Khalis. Second only to Jamiat in quality.

▨ Hezb-i-Islami breakaway faction; led by Gulbuddin Heckmatyar. Receives the most arms but has third best fighting strength.

☐ Jawzjani Uzbek mercenery army of 100,000; led by Abdul Rashid Dostam

☆ Ittehad-i-Islami: Led by Muhammad Afzal backed by Saudis.

Traditionalists

Harakat-i-Inqilib; led by Nabi Mohammedi. An inefficient group.

≡ National Islamic Front of Afghanistan; led by Pir Sayyed Ahmed Gailani; many religious followers.

★ National Front for the Rescue of Afghanistan; led by Sigbatullah Mojadedi, a monarchist and, in 1992, Interim President.

Minor Resistance Groups

▦ Shia Muslims. Much talk but little action.

☐ Other groups, either non-aligned or with mixed loyalties.

The Communists

▦ The People's Democratic Party of Afghanistan, with Soviet support. The PDA is an amalgam of the Parcham and Khalq factions. President Najibullah (Comrade Najib) is a Parchamite. The second most important figure is Muhammad Gulabzoi, a Khalqi and the Interior Minister. Parcham-Khalq enmity is notoriously bitter.

bring about military co-operation. There were unconfirmed reports that some of the reformist field commanders had been murdered. Other enlightened men have certainly lost their lives, including Professor Bahauddin Majrooh, Director of the Afghanistan Information Centre in Peshawar. Majrooh arranged an opinion poll to find out how ordinary Afghans wanted to see the conflict ended. His poll showed great support for the return to power of ex-king Muhammad Zahir Shah. This result angered Gulbuddin Hekmatyar, who sent gunmen to murder Majrooh.

It became clear in 1990 that the real struggle in Afghanistan was between Afghan nationalists and the Islamic fundamentalists. Their aims could hardly be more opposed. The fundamentalists want strong central control, the rearming of the country on a systematic modern basis, and the imposition of a tribal and Islamic ideological base on Islamic society.

The nationalists want a peaceful settlement, decentralisation of authority and administration, a demilitarised government and, above all, ethnic reconciliation. The fundamentalists would accept help only from Islamic nations; the nationalists want the world community to help them rebuild the broken country.

Three countries in particular were trying to gain control in Afghanistan in 1990 — Pakistan, Iran and Saudi Arabia. The Iranians had the support of the Shia-backed rebels in central and western Afghanistan. The Pakistanis wanted a pro-Pakistan swathe of territory in the Pathan tribal areas; and the Saudis were depending on their vast wealth and hoped to achieve their aim of dominating Afghanistan through General Tanai. At the end of 1990 several experienced Western observers in Afghanistan suggested that 1991 would be the 'year of decision' in the country.[2]

Najibullah's Nightmare — Superpower Co-operation

In the latter part of the 20th century no country's affairs stand in isolation. In particular, the affairs of the superpowers and of the other great powers affect those of scores of others. The government of Afghanistan learnt this lesson in 1991 and in a sharp and unexpected way.

In August, the news of the coup against President Gorbachev caused jubilation in Kabul. The old guard Communists confidently expected, at least for a few days, that with the despised peace-loving Gorbachev out of the way Moscow would revive its old anti-Western line. This would mean, they believed, fresh Soviet support for Kabul's war against the Mujahideen.

When the anti-Gorbachev coup crumbled, it became clear that the Soviet Union was in the throes of a democratic revolution and politicians in Kabul then feared that Moscow might cut its ties with Afghanistan. Najibullah and his colleagues were in despair and urgently sought reassurance from Moscow. It was not forthcoming and at all levels most requests for information about future Soviet intentions were ignored. The first blow was that the flow of Soviet wheat and fuel supplies slowed. In the third week of September Kabul was hit by a crisis. After two years of fruitless negotiations, the US and Soviet Union announced that in January 1992 they would stop supplying arms to the Mujahideen as well as to the DRA. This accord was announced in Moscow by

Secretary of State Baker and Soviet Foreign Minister Boris Pankin. It was intended to force the warring parties to the conference table and eventually manoeuvre them towards a UN cease-fire plan, an interim government and free elections.

Najibullah was taken by surprise but accepting the inevitable, turned the situation to his advantage. 'We consider it a turning point', he said. 'It is the beginning of a restoration of peace.' With military assistance from the Soviet Union at an end, he had decided that negotiation was more acceptable than continuing the war. No other war supplies were even remotely in sight, while the Mujahideen were still receiving weapons from Pakistan, Saudi Arabia and other Arab sources.

Baker and Pankin urged all countries involved in the war to co-operate in the arms embargo but few analysts believed that Pakistan's ISI, the main provider of arms to the Mujahideen, would comply. Najibullah, speaking to a foreign journalist, said:

> Pakistan talks of a political solution but it has a double policy. However, Pakistan will not in the end be able to evade fulfilling its commitments under the American–Soviet plan. The Baker–Pankin statement explains that the supply stoppage will take place from all sides to all sides.[3]

The DRA's military capacity will not be handicapped for some time, perhaps several years. It has enormous stockpiles of every kind. The danger for Najibullah is that his troops will see him as a leader without a future and change sides.[4]

Najibullah's situation was precarious even before the arms ban was announced. He could barely feed his government and party workers, all of whom depend on subsidised rations to survive.

The Mujahideen are in a correspondingly good position with arms stocks, which is why Gulbuddin Hekmatyar made a show of welcoming the US–Soviet arms ban. His group has the largest weapons stockpile among rebel organisations. While working on a plan to attack Kabul, he was hoping to persuade wavering Najibullah supporters to mount a coup in the capital.

Anticipating this move, Najibullah declared: 'Those dreaming of a coup should know that our forces have defended the country for 30 months without Soviet troops. They can do the same in the future.'

In the latter months of 1991 there was a build-up of Mujahideen forces in several parts of the country and plans were being made to attack six or more cities. All were bound to prove hard to take provided army morale remained solid.

Najibullah's Political Ploys

Najibullah quickly had a strategy in operation after the joint US–Soviet peace campaign was announced. He set out to convince the West that his Watan (Homeland) Party had abandoned Communism and endorsed democracy, capitalism and Islam. This all-embracing sympathy for opposing ideologies was spurious. By pretending to spread his support widely, Najibullah was hoping to win leaders of all political, religious and social groups to his cause.

When he announced his endorsement of capitalism and democracy he said that this entitled him and his party to a voice in any peace settlement. Simultaneously, he hoped to convince Muslim leaders that he himself was a devout Muslim — even though he was fighting Muslims. The ploy was not successful. He and his officials warned the West that if Hekmatyar came to power an Islamic fundamentalist movement would spread into Central Asia and start a new cold war, between the West and Islam.[5]

Najibullah told a press conference in Kabul that he would talk to the rebels without preconditions but that Hekmatyar remained the exception. Reconciliation with this man was impossible because 'he is too far away for me to extend my hand.'

The Mujahideen's response was that Najibullah must step aside, or at least give up control of the state security apparatus, before negotiations could begin. This is not even a possibility, for it would be tantamount to committing suicide. Diplomats in Kabul told me as early as September 1991 that Najibullah was cornered, with nowhere to run, but because of that very fact, this armed and determined man was dangerous. If the Mujahideen decided on a military push the only certain result was the loss of many more lives.

Just as the accord between Washington and Moscow to stop supplying arms to the belligerents was giving rise to cautious optimism about peace prospects, the rebels launched an offensive against the city of Gardez, 'the gateway to Kabul.' This attack, on 1 October 1991, was made by several thousand rebels, but in itself it was neither unexpected nor alarming. However, it was spearheaded by tanks from an unlikely source and this caused concern in Western capitals.

Saudi Arabia gave the Mujahideen several hundred Iraqi army tanks captured by the American and British forces in the Gulf War and turned over to the Saudis. The Saudis have supplied them to the rebels as part of their determined attempts to gain influence in Afghanistan when the war finally ends.

The Saudi move was seriously destabilising because Moscow had announced that if the rebels were reinforced and resupplied by the United States' allies — meaning Saudi Arabia and Pakistan — it might consider re-arming the DRA. The captured Iraqi tanks, all in good condition, were crewed in the Gardez offensive by Afghans who went to the Gulf as part of the coalition forces but who did not take part in the drive that threw the Iraqis out of Kuwait. The US State Department tried to halt the rebel offensive but without success.

Mujahideen forces around Gardez rocketed and shelled the defences for more than two weeks before beginning their ground attack. They closed the road to Kabul, thus threatening the capital itself but yet again it seemed that the Pakistan ISI had planned the offensive using tactics that have repeatedly failed.

Simultaneously with the Gardez operations, a Mujahideen peace campaign began. Sibghatullah Mojaddedi led a group of relatively moderate Resistance leaders to New York to meet the UN Secretary General, Javier Perez de Cuellar. The party included the monarchist leader of the National Islamic Front for Afghanistan (NIFA), Pir Gillani. NIFA is considered by diplomats in Peshawar as the only genuinely moderate party of the Resistance.

The Pakistani intelligence officers were intent on subverting the peace process and to this end they added several hardliners to the list of delegates to visit Moscow. They threatened to cut the flow of arms to Mujahideen commanders who refused to follow ISI guidelines when negotiating with US or Soviet officials.

The Islamic Dimension

While the Resistance or Opposition to the Kabul government is usually divided in the West into Traditionalist Moderate, Islamic Fundamentalist and Iran-based Factions, the simple fact is that all the groups are heavily influenced by Islamic attitudes, teaching and practice. The very fact that President Najibullah has consistently appealed to Islamic sentiment in his efforts to rouse support for his policies has shown the strength of Islam in Afghanistan.

The three 'traditionalist moderate' parties are, in reality, far from moderate and two, the National Islamic Front and the Islamic Revolutionary Movement, include the word Islam in their titles to emphasise the importance of Islamic tenets in their policies. The other, National Liberation Front, is led by a notable cleric, Sibghatullah Mojaddedi.

The four Islamic fundamentalist parties use the word Islamic in their titles. Two of the groups use the same name Islamic Party, and distinguish themselves only by reference to their respective leaders. They are: *Hizbi-Islami-Khalis* and *Hisbi-Islamic-Gulbuddin*. The other parties are the Islamic Union and the Islamic Society.

All six Iran-based parties are avowedly fundamentalist and Shia-dominated. They are: *Sazman-e-Nasr*, *Harakat-e-Islami*, *Pasdaran-e-Jehad*, *Hezbollah*, *Hehzat*, *Shoora-e-Ittefaq*.

Islamic passions, never far below the surface, are played upon by agents from all the Islamic nations involved in the war, notably by Saudi Arabia, Pakistan and Iran. All three maintain thousands of mullahs in the country as agitators and *agents provocateurs*.

Apart from their work among the groups opposing the Kabul government, Muslim extremists are carrying out a silent war of subversion and infiltration in the former Soviet Republics of Tadzhakistan, Uzbekistan and Turkmenistan. The long-term aim is to arouse the religious and nationalist passions of the 30 million Muslims north of the Afghanistan border. The work is inspired and financed from Saudi Arabia and Iran, though the two nations do not operate in concert.[6]

Relief Workers in Danger

In almost every edition of *War Annual* I have reported the difficulties of relief workers in bringing help to the victims of the war. Relief operations have always been dangerous but in 1991 *Médecins sans Frontières*, long noted for the courage of its teams, pulled out of the country. The risks faced by its doctors, nurses and their helpers were considered too great. One French doctor was murdered.

Four French Red Cross workers were held captive for 75 days by a Mujahideen commander. Another health worker was held hostage for 12 days against a ransom of $150,000 but was eventually released under pressure from other rebel leaders. Two American relief workers disappeared in the middle of the year and three Afghans attached to the Red Cross were murdered.

In September, a Red Cross official in Kabul said,: 'The situation is so serious that you cannot travel far down the road without the risk of being robbed or kidnapped by five different groups. We cannot send people into the field in these conditions.'

After Soviet troops withdrew in February 1989, independent agencies increased their voluntary efforts in anticipation of an early end to the fighting and the return of millions of the refugees. The organisations sent in hundreds of Western and Afghan specialists to clear minefields, rebuild irrigation systems and set up clinics.

Najibullah and some Mujahideen groups agreed to give the relief workers safe conduct but the aid agencies found that promises were not honoured, especially in areas where fighting continued. In the government town of Mazar-i-Sharif the government actually authorised the security forces to seize a wounded rebel under Red Cross care. After many difficulties, Red Cross officials pleaded with Najibullah himself to have the patient released. Soon after this, gunmen broke into the home of two Red Cross officials in the town and beat them up. The Red Cross, aware that the local military was taking revenge, suspended work in Mazar-i-Sharif.

Aid workers are no safer from the Mujahideen. At Faizabad, which is under government control, the Mujahideen promised to warn a UN group before they rocketed the town — but did not give the alert. Relief officials know that rebel leaders do not want the UN operating in government towns. The UN team pulled out of Faizabad. Unfortunately, this achieved the rebels' aim but the UN attitude is that, while its workers will take many risks, there must be a limit to the dangers to which they can be exposed.

Throughout 1991 and into 1992, it was evident that the problem was largely a matter of banditry. Rebel groups were turning to robbery and kidnapping to make up for the decline in military aid from Pakistan. Two Americans working for a Christian group called Global Partners were taken captive in July by the commander of a small Shia group in Ghazni. It demanded ransom in the form of a new road and several months later the Americans were still being held.

Despite the dangers, most relief agencies are still operating. Some say that the hazards are even greater proof that foreign relief work is necessary.

The 'Negative Symmetry' Accord

On 1 January 1992, an agreement signed by the US and what was then the Soviet Union came into effect. Known as the 'negative symmetry' accord, it binds both sides to cut arms supplies to their respective clients — the Russian side to the regime in Kabul and the Americans to the seven squabbling guerrilla groups. The agreement calls for the withdrawal of heavy weapons from the 'arena of conflict', cessation of hostilities leading to negotiations for peace, and the establishment of a transitional government.

The agreement called for much more than most observers expected could be delivered. However, the emergence of the central Asian republics, which had for 70 years been kept so firmly within the Soviet system, greatly alters the way in which Pakistan sees the Afghanistan conflict. For the first time in 13 years Pakistan's leaders had a genuine interest in helping to create peace in its neighbour, Afghanistan. This is because the new Asian republics open up vast new commercial opportunities for Pakistan. The nearest of them are only 16 hours by truck from the North-West Frontier and the route is through Afghanistan. Karachi, in Pakistan, is the republics' nearest port and it could benefit immeasurably from the revival of trade through Afghanistan.

While Pakistan is not a signatory to the negative symmetry accord, it acted throughout the war as distributor for American and Saudi arms going to the Mujahideen. Following the accord, Pakistani officials began to talk of the need for direct talks between the Mujahideen and the Kabul government. Deprived of their influence over the guerrillas because the arms supply was drying up, the Pakistanis were forced to make a rapid about-face. Prime Minister Nawaz Sharif and his army chief of staff, General Asif Nawaz Janjua, agreed on the policy shift.

As 1992 progressed, President Najibullah was also trying to build friendships in Central Asia. The former communists, who now call themselves 'democratic socialists,' have much more in common with Najibullah than with the guerrilla resistance groups. None of them is interested in having a fanatical Islamic regime in Kabul; they have trouble enough with their own Islamic movements.

Afghanistan has another problem. After years of war and foreign aid, Mujahideen commanders are clinging to the territory they control. Many have found the profits from drugs trafficking too profitable to contemplate giving them up. At the beginning of 1992, the United Nations Drug Control Programme estimated that 2,500 tonnes of opium would be grown in southern and eastern Afghanistan during the year. This figure suggests that the Afghan poppy crop would have a street value of more than $100 billion.

Helmand Province is under the control of Haji Muhammad Rasul and, on his instructions, 60 per cent of the province's cultivable area was sown with opium poppies. According to British mine-clearing teams who have worked in the area, opium is even grown in the minefields.

The Regime Collapses

After 13 years of indecisive warfare the Afghanistan conflict took a dramatic turn in May 1992 when the Mujahideen guerrillas virtually swept away the defenders of Kabul. For more than a decade rebels had been firing Soviet-made rockets into the city, trying to unnerve the Communist regime of Dr. Najibullah and suddenly there were signs that this was happening. Various guerrilla groups drew ever closer to Kabul, notably the alliance between Ahmad Shah Massoud of Jamiet-e-Islami, Sayed Mansour Naderi, the Ismaili chieftain whose militia control Baghlan Province, and General Abdul Rashid Dostam, leader of the wild Uzbek militia. With these three was General Abdul Mohmin, whose 70th Brigade mutinied against the government in January. Mohmin was working closely with General Dostam.

The key figure in the new developments was Massoud, known as 'The Lion of Panjshir', a Tajik from the north. He had become so influential and powerful that even the Afghanistan Foreign Minister, Abdul Wakil, left the capital to visit Massoud in his temporary headquarters. Wakil knew that the fall of Kabul was imminent and hoped to ensure that the city would not be destroyed. Najibullah sought refuge in the United Nations building and liberation came swiftly. The Mujahideen controlled Kabul and the war was technically at an end in the last week of April.

The first fighters to enter Kabul were supporters of Gulbuddin Hekmatyar, a bitter rival of Massoud, who had been frozen out of the conquering alliance. He at once began a new war against Massoud and his allies. Even as an interim president, Sibghatullah Mojaddedi, tried to form a new government Hakmatyar's rockets rained down on Kabul's southern suburbs and areas near the presidential compound. Guerrillas and militia loyal to Mojaddedi responded with heavy artillery and tank fire on Hekmatyar's positions. Massoud's men then slowly pushed Hekmatyar's infiltrators from the city.

A patchwork government of 51 Mujahideen leaders formally assumed power with Mojaddedi, who named Massoud as Defence Minister. Mojaddedi offered an amnesty to combatants but said that Hekmatyar was an 'aggressor' who would be punished under Islamic law if his forces did not cease firing. The following day Pakistan's Prime Minister, Mian Nawaz Sharif, and Prince Turki al Faisal, the head of Saudi Arabia's intelligence service, flew into Kabul to show support for the Mojaddedi government. The prince said, 'The Mujahideen victory has become a symbol of pride for the Islamic world'.

Afghanistan's new rulers will not be able to exercise real authority for a long time. Hekmatyar remained powerful and a Shia Islamic group with close ties to Iran controlled a large part of Kabul, demanding great representation in the government. Throughout Afghanistan rival guerrilla groups vie for power, often violently. Nevertheless, by the end of May streams of Afghan refugees were going home. According to relief agencies the refugees were gambling with their lives, because they face the risk of starvation, being crippled by landmines or being killed in the vendettas which rage between rival Mujahideen chieftains, even in the villages. The Najibullah phase of the war had ended, another phase was beginning.

The new phase in Afghanistan was a resumption of old enmities. The life and death of Abdul Mohmim, an army general who turned into a warlord, shows why Afghanistan cannot expect peace. Mohmim's 70th Brigade helped to capture and then hold Mazar-i-Sharif, the principal town in the north. At the time Mohmim, as a general, was loyal to the government but he deserted to the Mujahideen, thus setting off the chain of events that resulted in President Najibullah's downfall. Simultaneously the Uzbek militia, led by Abdul Rashid Dostam, turned against the government.

The final event in this sequence was the seizure of Kabul airport by the Uzbeks. This effectively prevented various Mujahideen governments from exercising control over the nation. How can any government actually govern when it down not 'own' the one international airport? President Burhanuddin, a shrewd operator, gradually bought off his opponents with money and promises, knowing that the time would come when he could make some decisive move. It came when Rabbani induced General Mohmim to change sides a second time. On New Year's Day 1994 Mohmim not only joined the government side; he attacked General Dostam's Uzbeks in Mazar-i-Sharif.

We do not know what reward Rabbani promised Mohmim but he did not live to receive it. He died in a helicopter crash in northern Afghanistan during the first week of January 1994. Helicopter fatalities are strangely commonplace among Afghan leaders.

Meanwhile the Hizbe Islamic leader, Gulbuddin Hekmatyar, had also been attacking the Uzbek militia in Kabul, by shelling and rocketing their positions. The Uzbeks did not leave but from Hekmatyar's point of view they did the next best thing. With their leader, Dostam, they came over to his side. The fighting was bloody in Kabul and Mazar-i-Sharif, with both sides using jet aircraft to bomb the other's positions. The innocent inhabitants, unable to escape, perished in their thousands.

The government forces did not simply allow the Uzbeks to have their own way but forced them to retreat to the ancient citadel of Bala Hissar within Kabul. The Uzbeks probably number no more than 4,000 but they are so heavily armed that no opposing force would attempt a frontal assault on them.

The scope for exercise of hatred is immense. General Dostam is dissatisfied with President Rabbani and even more with his main military commander, Ahmed Shah Massoud, both of whom are Tajiks. Hekmatyar, the most powerful military leader among the Pushtuns, the largest group, was in January 1994 withholding total support for the Uzbek militia. The northern alliance that had until then kept the Pushtuns out of power was falling apart and it suited the Hizbe Islamic leader to wait and see where his main advantage lay. For the ordinary Afghans the prospects were bleak.

Rabbani and his supporters were attempting to portray the struggle as one against diehard Communists, which it was not. Afghanistan had sunk into a terrible state of medieval treachery, made even worse by all the groups possessing modern weaponry. In January, many people were wishing that Najibullah, despite his great shortcomings, had remained in power. Believed to have taken sanctuary in the United Nations building in Kabul, Najibullah was scheming to revive his fortunes.

War planes from rival factions, together with artillery, continued to cause destruction and loss of life. Kabul's historic Blue Mosque was reduced to rubble and in 10 days of fighting at least 400 people were killed and 4,500 injured, according to the UN. Intermittent fighting was still going on in May 1994.

References

1. Diplomats in Peshawar and Islamabad have frequently told me of their frustration and exasperation with the Resistance leaders. 'It is impossible to debate any subject rationally with these men', said a German. 'They see approval of somebody else's idea as a sign of weakness.'
2. And so it proved. The year was decisive if only that the Americans and Soviets agreed on a policy for Afghanistan.
3. To Edward W. Desmond, *Time Magazine*, 23 September 1991.

4. An Indian diplomat in Kabul said; 'Psychology, not arms, is the weakness of any regime. Loyalties in Afghanistan have no basis in loyalty. Everybody in authority is self-serving and ready to switch allegiance.'
5. He spoke in alarmist language and tone but the scenario he anticipates in the event of a Hekmatyar triumph is valid enough. Hekmatyar is known as a man devoid of compromise, an admirer of the late Ayatollah Khomeini, who also abhorred compromise. Farid Mazdac, deputy secretary-general of the Watan Party, told me: 'If Afghanistan becomes an Islamic state the world can say goodbye to stability in Asia for a century.'
6. An official of the Soviet Foreign Ministry said that his government had complained to Saudi Arabia and Iran about the activities of their agents but that all protests have been ignored.

Major aspects of the Afghanistan war dealt with in *War Annual No. 5* include:
    The personality of Najibullah.
    The plight of refugees.
    Infiltration against the Soviet Union.
    Self-destruction tendencies among Mujahideen leaders.

Guerrilla Civil War in Angola

PEACE — PERHAPS
Background Summary

Angola became independent from Portugal in 1975 and at once a civil war began between the two main parties. The Popular Liberation Front of Angola (MPLA) formed the government and was opposed by the National Union for the Total Liberation of Angola (UNITA). The Soviet Union, seeking to dominate a country with strategic control over the sea routes around Africa, backed the MPLA and asked Cuba to send 13,000 troops to support MPLA's forces. In counter-balance, UNITA drew help from South Africa, which saw the opportunity to stop guerrillas of the South-West African People's Organisation (SWAPO) from using Angola as a base for attacks against South Africa. The United States, in order to counter Soviet moves, supplied arms to Dr. Jonas Savimbi, UNITA'S leader.

Of the several notable battles that took place, the most important were UNITA's defence of its base at Jamba early in 1988 and the battle of Cuito Cuanavale, in July–August that year, between the allied Cuban-MPLA army and UNITA–South African troops. It was a drawn encounter.

South African and Cuban troops agreed to withdraw from Angola following the Angola–Namibia Accord, signed at the end of 1988. Savimbi distrusted the Cuban promises but met the MPLA President Dos Santos and agreed on a ceasefire and peace talks. Fighting continued, but UNITA held 16 of the 19 provinces. By 1990, 500,000 Angolans were refugees in neighbouring countries and another 500,000 were displaced internally.

Two international events influenced the war. Namibia became independent, thus providing a home for the SWAPO guerrillas who had previously used Angola as their base. This denied South Africa any justification for keeping its troops in Angola, though it promised to maintain an arms supply to Savimbi. Secondly, in Europe the Warsaw Pact fell apart and, of even greater significance, this collapse had the Soviet Union's approval. President Fidel Castro of Cuba realised that Soviet expansionism in Africa was at an end and that he had no future in Africa. In the meantime, American aid for Savimbi was radically cut, a policy which influenced President Gorbachev and his ministers, who were seeking a reason for withdrawing from their Angolan commitments.

War and Peace in 1991

Persistent drought in Angola, as well as the armed conflict, spread starvation across vast areas of countryside in 1990. UNITA considered any food shipments organised by the MPLA government as military targets and the Angolan people were in a desperate situation, especially in the central province of

Kwanza Sul.

The UN organised relief efforts but they were hampered by poor donor support, weak organisation and lack of communication procedures between the opposing forces. The UN programme was abruptly stopped following a UNITA attack on a bridge in the southern province of Huila. The Angolan government, angry over this incident, accused UN workers of supplying military information to UNITA and suspended the UN's right to travel to UNITA areas. After such a setback, months elapsed before trust and communications were restored.

The Angolan drivers employed to take the relief trucks to distant destinations ran terrible risks. Both sides ambushed UN columns, mainly to establish and reinforce their authority. 'Inspections' of the relief trucks lasted for days and sometimes undisciplined units looted and destroyed them.

At the end of 1990, UN officials believed that the importance of the emergency programme lay as much in its symbolic value as in the food and medicine it provided. The government and UNITA appeared close to a settlement and the UN believed that if both sides agreed that civilians throughout Angola had the right to international aid, the peace process would be strengthened. This seems to have happened.

After 16 years and at least 350,000 deaths, the civil war ended on 31 May 1991. In Lisbon on that date President Dos Santos and Jonas Savimbi signed a peace pact. US Secretary of State James Baker and Soviet Foreign Minister Alexandr Bessmertnykh were the pact's brokers. The way was clear for 'immediate' establishment of multi-party democracy and elections, in 1992. The superpowers announced that they would no longer send arms to their respective proxies.

Angola had been a bottomless pit of expense and despair. The Soviet Union sent in 1,100 advisers, 50,000 Cuban troops and up to $1 billion a year in military aid for the MPLA. Washington's official financial assistance amounted to $60 million annually but diplomats in Luanda say that it was actually much more.[1] The US has never admitted to having sent 'advisers' to Jonas Savimbi. In fact, several hundred Americans worked for UNITA though they could not be accurately classed as 'advisers'. Savimbi was not the man to accept advice. The Americans were, in fact, military instructors.

US aid for UNITA caused friction in Congress and preoccupied four successive administrations, all of them unhappy about overseas entanglement after the agony of Vietnam but desperate to block the spread of Communism in the Third World: Presidents Bush and Gorbachev, both of them without personal commitment to the Angolan quarrel, were able to collaborate in the great effort to establish peace.

After Mozambique's renunciation of Marxism–Leninism in 1989 and Namibia's peaceful transition to democracy under former guerrilla leader Sam Nujoma, the Angolan settlement deprived South Africa of its much-used alarmist warning that Communism was taking over southern Africa.

Protecting the Peace

Within weeks of the pact being signed, experts from the US, Soviet Union, Britain, France and Portugal were working together to help MPLA and

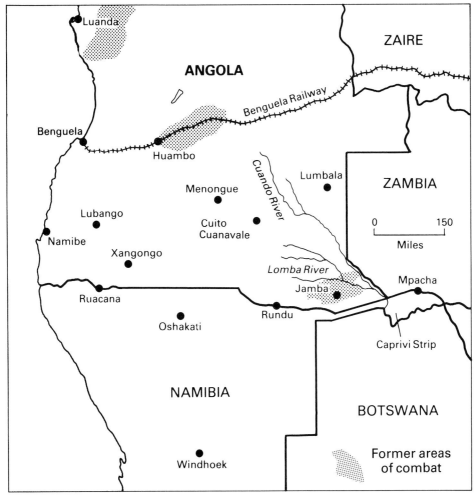

Angola: Areas Where Fighting Has Died Down

UNITA officers decide how to confine their troops to assembly points and lay the groundwork for the demobilisation of 200,000 fighters and the integration of the rest into a single 40,000 strong national army. The release of prisoners-of-war began at the end of July.

Tremendous difficulties lay ahead. Many people complained that because the demobilisation and integration of the armies would be complete just before the elections, civilians could still be intimidated to vote for the MPLA. Registration of new parties must meet government requirements. Civilians wanting to sign up to support the political parties had to produce many photocopies of documents and stamps which were virtually unobtainable.

The MPLA has the advantage in pre-election campaigning. It has a revenue of $3 billion, mostly from oil, and it controls the media and the massive bureaucracy.[2] Military officers and party officials have been given access to foreign exchange to buy cars and ensure a high living standard in the event of an MPLA defeat. Not that the MPLA has anticipated defeat. Under the peace deal, UNITA agreed under pressure to allow the government to extend its control throughout the country. It was obvious to foreign observers in the country that this gave MPLA a tremendous electoral advantage.

However, evidence of UNITA's appeal and strength is not hard to find in certain areas, particularly on the Planalto, Angola's high plateau. The organisation's military skill can be seen in the many hundreds of ambushed MPLA vehicles. Cuban troops helped the Angolan army to drive UNITA from the city of Huambo in 1975 but UNITA constantly harassed the army units. Huambo province has one of the country's highest concentrations of anti-personnel mine victims. Much killing took place after the ceasefire but order was restored by Brigadier David Wenda, UNITA's commander in Huambo Province, and the small unit he bought in from the bush to act as administrators.

Late in July 1991, UNITA's failure to release government prisoners-of-war threatened to undermine the great gains achieved under the ceasefire. The crux of the problem lay in UNITA's inability to explain the apparent disappearance of about 180 government soldiers held captive by the rebels in 1990. In October 1990, the Red Cross interviewed 202 captured government soldiers at Jamba, UNITA's HQ. But when UNITA presented a list of captured soldiers early in July only 29 of those interviewed appeared. The government has claimed that UNITA should be holding 1,192 POWs, while the rebels have said they have 371.

Angola's peace is decidedly not the result of military victory or reconciliation. Its sole precipitating cause is the enfeeblement of the belligerents following the withdrawal of foreign support. MPLA and UNITA independently decided that they could not win control of Angola by force when South Africa dropped its support for UNITA and MPLA lost its Cuban troops.

A return of civil war remains a possibility and, in my opinion, a probability. An electoral defeat for Savimbi would send him back to the bush and reliance on his carefully secured weapon stocks.

Savimbi as a Military/Political Leader

Savimbi is revered and hated in Angola in equal measure, praised for being a liberator, condemned as a butcher who caused the long war. For those who view Savimbi as an heroic figure who stood against the power of the Soviet Union he is, naturally, a democrat. The large number of efficient, educated people who work for him acknowledge that he is an autocrat. His many enemies say that he is merely a despot who knows how to sell himself to the West.

Certainly, Savimbi understands the power of clever public relations and as he is a linguist — Portuguese, English, French, African languages and some German — he has no difficulty in expressing himself in meetings around the world. Among African leaders he stands out because of his air of authority and dynamism and, unlike most of the others, he is up-to-date in world affairs.

As a military leader, Savimbi is also outstanding; some experts consider him the best guerrilla leader of the century. All foreigners who have visited his HQ in the Jamba bush — in effect, his capital — come away impressed by Savimbi. He leads a disciplined Spartan life and expects the same from his men, of all ranks. Some visitors have described Jamba as a great boy scout camp, where honesty of purpose and decency of life are preached.

Essentially a field commander, Savimbi directs most of UNITA's operations in person and has even been known to be present at small-scale ambushes. Nevertheless he can delegate authority and is a shrewd judge of men who have leadership potential. Dressed in camouflage uniform and a black beret, he wears a pearl-handled revolver and a belt of cartridges, and carries a cane. Everything he does is for calculated effect.[3]

For visitors he considers sufficiently important — and that is most of them — Savimbi orders a military display. This is partly to demonstrate that his men are not mere bush bandits but trained troops who are as proficient on parade as they are in action. He also knows the importance of *esprit de corps* and in his long speeches he never fails to praise his troops. Even more, he always has a word for their families and the sacrifices they have made in devoting themselves to 'the struggle'.

The intelligent Savimbi knows that most of his visitors will also be visiting Luanda, a sleepy, seedy and corrupt city. Any comparison between Jamba and Luanda must favour Jamba. President Dos Santos is lack-lustre and pedestrian compared to Savimbi.

As a resistance leader, Savimbi has sound credentials. His grandfather fought the Portuguese in 1902 and as a result lost his chieftainship, a humiliating punishment. His father was also a freedom fighter. Aware of the value of an education, Savimbi begged a headmaster to give him a place and having completed a three-year course in two years, he won a medical scholarship to Portugal. After working with liberation groups for seven years outside Angola, he formed his own movement, UNITA. He was 29 years of age.

Always impressed with Mao's Long March and the type of socialism he built on peasants rather than urban workers, Savimbi cultivated the Chinese who gave him weapons and training. Nevertheless, by the time of the rebellion against Portugal, UNITA had won no military reputation.

UNITA, together with the Soviet-sponsored MPLA and the CIA-backed FNLA — National Front for the Liberation of Angola — (the CIA also helped

UNITA) signed the Alvor Accord of May 1990, which called for free elections. The Accord collapsed and foreign intervention increased. The Cubans rushed in men and arms to help the MPLA. The South Africans, in the mistaken belief that they had American support, sent a force to stiffen UNITA and headed in strength for Luanda. The Cubans were too strong for the South Africans and the US decided not to confront the Soviet Union in Angola. This disastrous military–political reverse forced the South Africans to withdraw from Angola. UNITA, too, was driven out and the MPLA, with its massive Soviet–Cuban backing, took over the country.

The setback might have finished many an African leader but Savimbi rallied his followers, restored their morale and built up a base in the south-east of the country. In 1980 UNITA troops were engaged in persistent guerrilla activity in the central highlands and frequently sabotaged the Benguela railway. The next eight years were bloodily successful for Savimbi and his units developed the capacity to operate a thousand miles from base. They planted belts of antipersonnel mines around innumerable villages and thus killed and maimed many thousands of civilians. He is now hated for this tactic which he justifies on the grounds of 'military necessity'. He has never explained the benefits to be derived from sealing off villages with minefields and, in fact, few MPLA soldiers were victims of the mines. They simply cleared the roads of mines and freely moved in and out of settlements.

The MPLA launched a major offensive against UNITA in 1988 and, with Savimbi's support, the South African Defence Forces counter-attacked. The confrontation at Cuito Cuanavale was the biggest in Africa since the end of the Second World War.[4] The South Africans were outmanoeuvred in 1989 when the Cuban army moved south to the Namibian border and the South Africans pulled back out of range of Cuban air attack.

Savimbi's power base is the Ovibundu people, who live in the central highlands. More than a third of Angola's population, they are antagonistic towards the plains and coastal people, who for generations have regarded themselves as superior to the Ovibundu. The Ovibundu's resentment can be seen from the slogans which Savimbi gave to UNITA — Socialism, Negritude and Democracy. Considering that Savimbi has for so long depended on white South Africa for survival, the Negritude slogan is ironic. The readiness to ally himself with racist South Africa and still claim that he champions negritude shows Savimbi at his pragmatic best and indicates that he is prepared to compromise to achieve his ends. But not endlessly. He wants supreme power and will not compromise this ambition. If democratic elections do not deliver it to him Savimbi will resort to other means.

Threats of New Violence

Both sides to the conflict were due to start demobilising their forces at the end of March 1992 but Savimbi showed little interest in keeping to the timetable. During the last week of March, he broadcast over UNITA radio an allegation that the MPLA was planning to launch an attack on UNITA and announced that he had fled to Jamba, his bush headquarters, because he feared a plot to kill him. At the same time, he ordered his soldiers at assembly points all over the country to 'stay on the alert.'

At the same time, there were reports of defections, disappearances and executions within UNITA. Diplomats in Luanda suggested that Savimbi was prepared to restart the civil war in order to regain his weakening control of UNITA. It seems probable that he had members of his own high command executed.

Savimbi's trusted general, Miguel N'Zau Puna, who was also UNITA's interior spokesman, and Tony de Costa Fernandes, the foreign affairs spokesman, defected in February 1992 and a month later appeared in Paris. There they accused Savimbi of committing crimes, including the execution of children, after the peace agreement with MPLA was signed. They also accused him of executing Wilson dos Santos and Tito Chigundji, the fomer UNITA representatives in Washington.

In April, most ambassadors in Luanda hoped that the elections would go ahead as planned in September 1992. Few believed Savimbi's claims that the MPLA would like to break the accord or that its demoralised conscript army could resume the war that brought devastation to the country.

Defying the Election Result

The elections did proceed. They were supervised by UN Observers, who pronounced them to be fair and honest. José Eduardo dos Santos won the elections but Jonas Savimbi and UNITA refused to accept defeat. This was entirely predictable. The war continued as if the elections had never been held and by August 1993 the town of Malange, with a population of 400,000, had become a main focal point in the conflict.

The MPLA army maintained a tenuous hold on the city and its fortifications, while hundreds of civilians were killed or mutilated when they ventured out into the belt of landmines surrounding the town in search of food.

Fighting around the town is perceived as a test of strength. Early in January 1994 four members of the US Congress, which had formerly backed UNITA with arms, were forced to abort a visit after UNITA shelled the town. In February UNITA gunners opened up again, wounding scores of people and killing as she slept a nurse who worked for the Irish charity, Concern.

Abuses and robberies by government soldiers continue, according to Angolan civilians and aid workers. In February 1994 a group of soldiers hijacked a truck hired by Concern. The provincial government has problems too. The governor of Malange, Flavio Fernandez, has been accused of selling food for profit.

Angolan aid workers say that aid that arrives very early in the morning or late at night is routinely looted at the airport. 'The level of theft by soldiers is just terrible', Miss Fernandes, the Caritas representative said. 'The only way to end these abuses is to stop the war.'

The war is unlikely to end. Both President dos Santos and Jonas Savimbi fear for their safety and prestige should they concede too much to the other side. Negotiations began again in December 1993 but by May 1994 were deadlocked. In the meantime tens of thousands of emaciated Angolans shuffled into feeding centres in Malange each morning. A massive food programme has thwarted a famine that was killing more than 100 people a day in January 1994.

Up to 140 tonnes of food a day reaches Malange, all by air because the town is besieged by Savimbi's troops. Fighting around the city is sporadic with each side provoking the other.

The only reason that the belligerents are not more violent is that they have lost their respective great-power backers. The old Eastern European bloc can no longer supply MPLA and the US gives UNITA very little. Even the Cubans, who had vigorously and generously aided President dos Santos, are no longer heavily involved in Angola although for the sake of prestige they maintain 'advisers' in Angola.

References

1. A Zambian official claims that he has seen figures which prove that between 1980 and 1988 the annual amount averaged $300 million, including the cost of arms.
2. Corruption in Angola, and in Luanda in particular, is the main talking point in diplomatic and press circles. According to a Swiss official it is worse than in Cairo, which is notorious for corruption and bribery.
3. Having seen Savimbi in action on several occasions, I am impressed by his professionalism as a political performer. He trained his staff as early as 1980 to refer to him as 'the President' when conversing with foreign officials and journalists so that his future status would be taken for granted.
4. Several writers have picked up and repeated the statement that the battle of Cuito Cuanavale was the biggest in Africa since that of El Alamein in October 1942. This is not so. The battle for the Mareth Line and other actions in Tunisia and Morocco were larger in scope than Cuito Cuanavale.

Bangladesh War of Genocide

TIME RUNS OUT FOR THE RESISTANCE

Background Summary

When India and Pakistan became independent from Britain in 1947, the chiefs of Buddhist tribes living in the Chittagong Hill Tracts hoped for recognition as a native state or as part of a confederation with tribal areas of north-east India. Instead, the British incorporated the Hill Tracts into Pakistan, a Muslim state. The 600,000 Buddhists of the Tracts, in a total of 32 tribes, feared for their future under expansionist Islam.

Until 1971 they came under only desultory pressure but everything changed after the bloody civil war between East Pakistan and West Pakistan. East Pakistan became Bangladesh and the new government at once began to drive out the tribes in order to give their lands to Muslim settlers from Bengal.

Alarmed, the Buddhists formed a 'self-defence association', the *Jana Sanghati Samity* (JSS). The largest tribe, the Chakmas, created a military wing, the *Shanti Bahini*. The Bangladeshi government ridiculed this force, since the Buddhists have no military tradition, but the guerrillas showed so much natural ability that for 13 years they frustrated government efforts to crush them.

In 1984 Major General Noor Uddin Khan was ordered to find 'a permanent solution to the Hill Tracts problem'. His solution was terrorism, deportation and genocide. The terror tactics of his 24th Division, the Bengal Tigers, could not eradicate village support for the 5,000 *Shanti Bahini*. The General's offer to the Buddhists that they could stay in the Hill Tracts if they converted to Islam was also fruitless.

Uddin Khan resorted to systematic atrocities to bring the guerrillas out of the jungle. The tactic was to impose a regime of torture, rape and execution on one village at a time. In effect, holy war or *jihad* was declared against the Buddhists as 'infidels and unbelievers'. The objective was creeping genocide. Still the *Shanti Bahini* held out, although supplies of money and weapons from the Soviet Union and China dried up.

The government caught no more than 10 guerrillas in 1990, largely because they hid in tunnels during intensive army dragnet sweeps. The Bangladeshi government approved Uddin Khan's scheme to bring in gangs of armed and vicious Muslim zealots to hunt the *Shanti Bahini* but they were targeted by the guerrillas who killed more than 300 of them in 1990.

The Chakmas and the other Buddhist tribes of the Hill Tracts are not a popular cause and attract no international concern because they have no outside spokesmen to represent them. Only Amnesty International and the Anti-Slavery Association have expressed sympathy. When Rajiv Gandhi was

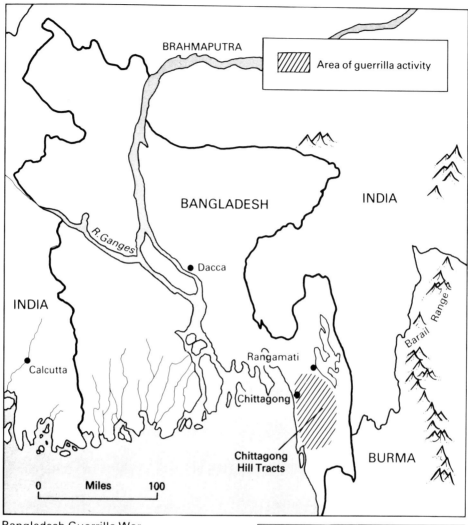

Bangladesh Guerrilla War

defeated in the Indian elections of 1990, the Chakmas sent representatives to New Delhi to plead with the new Prime Minister, V. S. Singh, for support, but he was too briefly in power to help them.

The War in 1991

The downfall of President Hussein Muhammad Ershad in December 1990 and his replacement by a woman ruler brought no respite to the Chakmas and other oppressed Buddhists. Unusually, during the monsoon period of May–July the army continued its campaign against them despite the difficulties of maintaining operations in the extreme wet. The theory seems to have been that the *Shanti Bahini* would be forced to take shelter and thus could more easily be trapped. According to some UN relief workers who penetrated the Hill Tracts in August, the reverse happened. As Bangladeshi troops moved slowly along the jungle trails, under appalling conditions, they were ambushed by the guerrillas. The troops could not pursue their hit-and-run enemy and suffered 'numerous casualties'.

During the monsoon the coastal areas of Bangladesh were hit by typhoons, floods and tidal waves and scores of thousands of lives were lost. American warships and troop transports returning home from the Gulf were diverted to help the stricken Bangladeshis. At this time unidentified groups urged the US administration to find some way of reaching the Chakmas with humanitarian aid. The request was refused on the grounds that this would be interference in the affairs of a friendly state without its permission.

A confidential report out of the Hill Tracts in September 1991 said that 'friends of the Chakmas' managed to raid Bangladesh army supply depots during the chaos caused by the floods and steal badly-needed weapons and ammunition. The army was completely unaware of these thefts. The *Shanti Bahini* was ready for the army when it resumed normal operations after the monsoon.

The same report refers to the 'exhaustion of spirit' among the Buddhist peoples, who find themselves in an ever shrinking area. Their social infrastructure is weakening and the total collapse of the people's will to resist may be only a few years off.

The War in 1992-94

Despite the prediction that the Buddhists would give up their struggle they have persisted in their efforts. European travellers late in 1993 said that they possessed 'an amazing reservoir' of courage and that this was strengthened by the conviction that sooner or later the outside world would come to their aid. One of the European journalists who penetrated into the Hill Tracts tried to explain that if the United Nations would not help the suffering people of East Timor under Indonesian occupation and oppression, they were not likely to help the Buddhists of the Hill Tracts. Apparently the Buddhists are incapable of accepting this. The Western Christians will help them when their present preoccupations come to an end, they say. However, in 1994 there was no sign that the UN or any grouping of nations would go to the aid of the Buddhists.

The causes of the East Timorese, the Burmese and other oppressed peoples survive because of the oxygen of outside publicity. The Bangladeshi government has continued to prevent foreigners from entering the war area and the few who manage to do so are soon discovered and roughly expelled. Amnesty International continues to recognise the conflict between the Bangladeshi government and the *Shanti Bahini* as a war.

Burma (Myanmar) Guerrilla War

THE KARENS' FIGHT FOR FREEDOM
Background Summary

The Karen National Liberation Army (KNLA) has been fighting for a homeland since 1949, together with guerrillas from Kachin, Kayan, Shan, Arakan, Mon, Naga, Kerreni and other tribes. Collectively, they comprise about 25 per cent of the 45 million people of Burma. The Karens are by far the largest group, with three million, but the KNLA has never had more than 5,000 fighting members.

Their claim for independence is well based. They fought with the British army against the invading Japanese during the years 1941–45 and were promised a homeland after the war. The British broke that promise. The fighting men and women of the ethnic minorities face the formidable defence forces of Burma — 300,000 in the armed forces, 40,000 of the Police Force and 35,000 of the People's Militia.

The Karens began their fight as guerrillas but they became regular soldiers and are organised as an army. Their commander is one of the longest serving soldiers in the world, General By Mya, who fought with the British during the Second World War. He is also president of the National Democratic Front Alliance (NDFA), the organisational body for the other independence groups. The KNLA's battlefield is the hill and jungle region of eastern Burma on the Thai border, while other groups are fighting in the Irrawaddy delta and in the mountainous north.

In 1988 the Burmese army called its operations against the Karens 'The Campaign of the Four Cuts'. They planned to cut the Karens' trade routes; cut off outside aid; to cut off one rebel group from another; and to cut off the rebels' heads. The 'Four Cuts' label has stuck.

Burma has suffered continuously from oppressive government since the Burma Socialist Programme Party (BSPP) imposed one-party rule in 1962. In August 1988 Maung Maung became President and promised multi-party elections, but because he did not set a date for them demonstrations continued. The national student movement rose against the government, to be joined by three leading dissidents — retired generals Aung Gyi and Tin Do and Aung San Suu Kyi, daughter of Burma's greatest nationalist hero.

General Saw Maung, Burma's Minister of Defence and chief of the armed forces, ousted Maung Maung and proclaimed himself President; however the most influential person has for long been the despotic Ne Win, former BSPP chairman.

To focus attention on a 'foreign' enemy and thus diffuse internal opposition, the army, whose leaders back the BSPP, began a new campaign against the Karens. Five of their bases fell but Kaw Moo Ra, in a curve of the Moei River that juts into Thailand, held out.

Meanwhile, the hopes of most ordinary Burmese rested on Aung San Suu Kyi. Leader of the National League of Democracy (NLD), she became by far the most popular Burmese politician. Afraid of her prestige, the military junta placed her under house-arrest in July 1989. General Tin Do suffered the same fate and the NLD was without leaders. The government, actually the State Law and Order Restoration Council (SLORC), changed the country's name from Burma to the Union of Myanmar.

In the national elections of 27 May 1990 the NLD won 72 per cent of the vote and 392 of the 485 available seats in the new National Assembly. The vote, in effect one for Aung San Suu Kyi herself, was a humiliating defeat for the SLORC and its military backers but they did not relinquish power. Aung San Suu Kyi was kept under house-arrest and told she would only be released if she went to Britain to live or renounced politics and took up a literary life.

Despite memories of the brutal crushing of the 1988 uprising, some protests took place. On 19 July 1990 small demonstrations called for the release of Aung San Suu Kyi and on 8 August, the second anniversary of the killing of demonstrators in Ragoon, at least two students and two Buddhist monks were shot dead by troops in Mandalay, Burma's second city. In reaction to the killings and the government's abuse of human rights in general, around 50,000 of Burma's 300,000 monks refused to administer religious services to the military and police and their families. In Burma, where the bulk of the population are deeply religious and monks are held in high esteem, the boycott was regarded as virtually an excommunication.

The monks' protest gathered strength and on 18 October Saw Maung issued an ultimatum to them to end their boycott in two days. When they refused, some monks' organisations were banned and military tribunals were given the power to impose harsh penalties on monks, including the death sentence. Many monasteries were surrounded, hundreds were raided and monks were arrested. Repeating its allegations against the NLD, the SLORC claimed that the monks' boycott was part of the Communist plot.

By 24 October the monks' protest was crushed. In a speech on 2 November Saw Maung made it clear that no form of political opposition from monks or other religious groups would be permitted in the future. The SLORC intensified its efforts to eliminate what remained of political opposition and on 23 October NLD party offices throughout the country were raided and closed and more leaders were arrested. Some are known to have been beaten to death.

Because of the election and its aftermath, the campaign against the Karens became less intense. Also, in June 1990 the new commander-in-chief of the Royal Thai Army, General Suchinda Kraprayoon, warned the Burmese army against making incursions into Thailand in order to attack the KNLA. This *de facto* support helped the Karens but they suffered badly when the army resumed its campaigns. By now the resistance groups were amalgamated into the Democratic Alliance of Burma, with a total of 15,000 troops, but about 1,000 lost their lives in the period June 1989 – June 1990.

The War in 1991

In April 1991 the military junta forced the NLD to drop Aung San Suu Kyi from the party leadership by threats, coercion, bribery and, in some cases, by beatings. This was a clever move because it meant the end of the three-year struggle against the dictatorship, and was the final step in the ruthless consolidation of military power.[1] She was replaced by U Lwin, a little-known political figure who could be easily manipulated. The announcement came one day after Burma's deputy army chief, General Than Shwe, was quoted as saying that there was no possibility of the army handing power to a civilian government. 'We cannot find any organisation that can govern the country in a peaceful and stable manner', he said.

Students had maintained occasional protests but as the universities remained forcibly closed their ability to organise themselves was limited. With systematic arrests of individuals and widespread intimidation of teachers and parents, the protests died out. In 1985 the SLORC had started to resettle many people from Rangoon, Mandalay and other cities in 'new towns', actually little more than barracks settlements where they could be easily controlled and 're-educated'. In mid-1991 this process was given new impetus so that by the end of the year up to one million people had been relocated.[2]

Dr. Sein Win, chairman of the National Party for Democracy, found refuge in the area controlled by the KNLA. The government sees Sein Win as a threat to its authority because he is a nephew of Aung San Suu Kyi, the nephew of her elder brother, who had been assassinated in 1947. At Manerplaw, the KNLA's HQ, Sein Win and a number of NLD elected representatives formed the National Coalition Government of the Union of Burma (NCGUB) on 18 December 1990. They are supported by the Democratic Alliance of Burma, a collective of the ethnic minority groups. All have stated that they will end the insurgency campaign once the SLORC hands over power to the elected representatives of the Burmese people.

The importance of all these developments is that the Karens' fight for freedom became inextricably linked to the general resistance against the oppressive military dictatorship. The Karens and the KNLA were no longer fighting alone, even if the new NCGUB was in no position to help them militarily.

Alarmed by growing popular support for the KNLA and other guerrilla groups, the government still further intensified its brutal actions against its political enemies. On 13 August 1991, 25 members of parliament were among 48 leading figures in the NLD sentenced to long prison terms for alleged treason and for 'attempting to establish an alternative government'. One of the four women MPs given 25-year sentences is a well-known Rangoon lawyer, Daw San San. She also happens to be an associate of Aung San Suu Kyi, the real reason for her victimisation. Another woman who was a member of Aung San Suu Kyi's personal staff was also given a 25-year sentence.

In mid-September 1991, Lieutenant General Aung Ye Kyaw told officials in the northern city of Mandalay to expect military control of Burma for a long time. 'We cannot say for certain how long we will be in charge of state administration', he said. 'It may be five years or ten.' He stressed that a

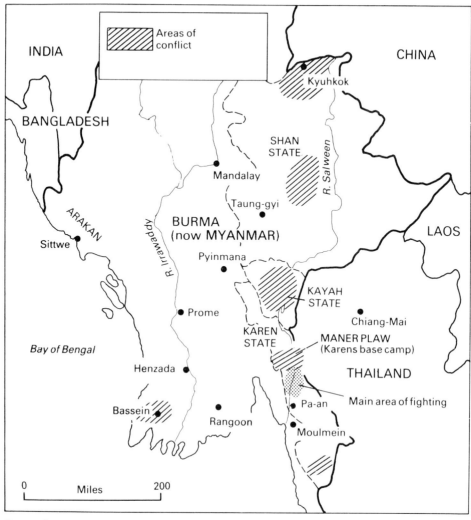

Burma (Myanmar) Guerrilla War

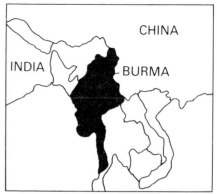

constitution had to be in place before the SLORC would transfer power. But everybody knows that the constitution which the SLORC has in mind will not favour democratic institutions.

Throughout 1991, the KNLA maintained the strength to attack the Burmese army. Virtually besieged, it sent innumerable patrols, some as large as 200-strong, through enemy lines to blow up dumps and to steal supplies. Recruits managed to penetrate the army cordon to reach Manerplaw in order to enlist in the KNLA.

The army has the manpower and military strength to launch a major offensive against the KNLA and the entire resistance organisation but the High Command estimates that its losses would be heavy. The generals believe that a defeat at the hands of the KNLA or even a part-victory might encourage an uprising throughout Burma.

The army has élite, well-trained units but they lack the confidence to confront the superb KNLA troops, who might well be the most proficient jungle fighters in the world.[3]

The Western Response

Phone Myint, Minister of Home and Religious Affairs and Secretary of the National Intelligence Bureau, warned representatives of various political parties: 'Foreign countries are watching us so that they can enslave and subjugate us. We do not trust the foreigners, not one bit. Nor do we think highly of them.' Phone Myint did not name the countries eager to 'enslave' the Burmese but China is the only country in a position to invade Burma and it has no interest in doing so. He may have been thinking of economic domination but there seems to be no obvious candidate. His comments, like those of other SLORC leaders, indicate raging xenophobia.

Western countries have urged the Burmese government to release political prisoners and respect the people's fundamental freedoms. The freezing of most donors' bilateral aid programmes originated when SLORC seized power and remains in force. In July 1990 when the SLORC's intention to renege on the results of the election became clear, many countries called on the government for an early transfer of power. On 7 September 1990 the British government condemned the arrests of NLD leaders as 'another blow to the cause of democracy in Burma'. Later that month, the 12 EC states led an 18-nation protest — with Australia, Canada, Japan, New Zealand, Sweden and the US — against the SLORC's human rights violations.

In January 1991, another EC statement emphasised that development aid programmes would remain suspended until the situation in Burma improved. On 9 July the EC agreed on a formal embargo of all arms sales to Burma and on 10 July in Strasbourg, Aung San Suu Kyi's son accepted the European Parliament's 1990 Sakharov Human Rights Prize on his mother's behalf.

On 30 May, at a meeting of foreign ministers of the EC and the Association of South-East Asian Nations (ASEAN), the British Foreign Secretary, Douglas Hurd, urged the ASEAN countries to exert influence on the Burmese authorities 'to adopt a more human approach to government'. He said he was concerned that a poor country, not subject to any external threat, should spend such a high proportion of its income on weaponry. The army increased in size from 190,000 in 1988 to an estimated 300,000 in 1991 and is still growing. Mr. Hurd said:

> The policies of the military authorities have maintained a potentially rich country in a state of unnecessary and artificial poverty, deterring trade, foreign aid and investment. Promotion of good government worldwide is the responsibility of the whole international community.[4]

On 10 September, Amnesty International issued another in its series of reports about conditions in Burma. It accused the military of widespread violations against ethnic minority groups, ranging from summary executions to conscripting people to clear minefields laid by the KNLA and other resistance groups. The human rights organisation said that abuse against the Karen, Mon and Indian communities occurred simply because the military had the power to perform them with impunity.

The Amnesty report stated that its own staff had held interviews along the Thai-Burma border with Burmese who had been victims of human rights violations or personally knew victims. Many of these interviews referred to murders and incidents of ill-treatment, sometimes resulting in death, the report said. 'Members of ethnic and religious minorities are seized as porters or used to clear mines.'

Shan State's Rebels

During 1991, the narcotics warlord and self-styled freedom fighter Khun Sa, took control of Shan State's rebel political council in eastern Burma. Khun Sa, believed to be the world's biggest supplier of heroin, was appointed president of the Tai Revolutionary Council, succeeding Mo Hang, a veteran rebel who had died of cancer.

Khun Sa, known to his many enemies as the 'Prince of Death', has long been commander-in-chief of the Maung Tai Army (MTA) which operates in Shan State. He has several times insisted that MTA is involved in the heroin trade only as a means of funding its fight for autonomy from Rangoon. He is the most notorious of several warlords who finance private armies through the production and trafficking of heroin in 'the golden triangle,' which straddles northern Thailand, Burma and Laos.

In theory, the revolutionary council controls the MTA but Khun Sa has always exerted the greatest influence, despite Mo Hang's political seniority. Diplomats who have travelled in the eastern areas of Burma say that the Shan certainly needed a new leader to revive their spirit to resist government oppression but that he must be a more popular man than Khun Sa.

A Shan official who broke with Khun Sa and fled to the Karens for protection accused him of exploiting the people and sacrificing hundreds of Shan lives in pointless battles against the Wa, a rival ethnic group. 'At least 800 young Shan were killed when Khun Sa fought against the Wa this year', he said in September 1991. 'He is a drug dealer without genuine political aims. We Shan should not have to sacrifice our lives for him.'[5] The US Drug

Enforcement Agency says that Khun Sa is wanted in the US on heroin trafficking charges.

It is known that four Shan State elders declined the position of president because they thought it was useless to compete against Khun Sa, who already held military and political control of the movement. According to British intelligence assessments, Khun Sa is a formidable military commander but they say his motivation is not liberation for his people but the protection of his narcotics empire.

Crisis Succeeds Crisis

During 1992 Burma was the scene of increased military violence despite several efforts by other nations to bring order and democracy to the country. The rapidity and intensity of the developments can best be shown in chronological order.

15 December 1991   Malaysia's Deputy Foreign Minister, Abdullah Fadzil Che Wan, said that the Association of South-East Asian Nations (ASEAN) would meet to discuss ways of restoring democracy in Burma. Japan announced that it would send the Deputy Minister for Foreign Affairs, Kukihiko Saito, to Burma to press the military to hand over power to the duly elected civilians whose leaders was Aung San Suu Kyi. On the same day, troops and riot police entered Rangoon University Campus, arrested 1,000 students at gunpoint and closed down all the colleges.

21 December 1991   Burmese border guards raided a Bangladeshi border post, killing one guard and wounding 12 others. At the same time the Burmese army massed troops along the 50-mile Bangladesh border claiming that this military activity was 'defensive.' Burma's ambassador in Dhaka, Soe Myint, said 'The build-up is in response to the growing insurgency along the border.' The so-called 'insurgency' was, in fact, the flight from Burma of tens of thousands of Burma's minority Muslim population, following atrocities committed against them.

6 January 1992   The Bangladeshi military discovered that Burma had built five helicopter pads and was reactivating a military airfield near the border town of Mongdu and, as a result, the armed forces were put on full alert. Burmese air force planes had already violated Bangladeshi air space. On the same day, the Rohingya Solidarity Organisation — the armed wing of the 50,000 Rohingya Muslim refugees who had fled their homes in Burma — announced that it had recruited more men in preparation to fight the Burmese army.

11 January 1992   The Burmese government yet again stressed the need for 'racial purity' among the ethnic Burmese majority and explained that it was protecting Buddhism by ridding the country of Muslims. The military rulers claimed that the Muslims had entered the country illegally from Bangladesh. (In 1978, a military purge in northern Arakan forced an estimated 200,000 Muslims to flee into Bangladesh, where they live in abject poverty as refugees).

15 January 1992   About 2,000 Burmese infantry attacked the KNLA's positions around the village of Malehtta, 20 miles north of the rebels' HQ at Manerplaw. The village was defended at the time by about 600 men from groups within the Democratic Alliance of Burma. Six bombers flew sorties against Malehtta, almost destroying it. In the subsequent infantry assault, the Burmese are reported to have lost 180 killed, with others wounded. The guerrillas held onto their fixed positions.

23 January 1992 Western Intelligence sources revealed that the eight Burmese brigades, operating in the Bandarban district on the frontier of south-eastern Bangladesh, had mined a 70-mile stretch of border.

1 February 1992 Australia's ambassador to Burma, Mr. Geoff Allen, met Burma's Foreign Minister, U Ohn Gyaw, in Rangoon, and again asked that power be handed to the opposition party which had democratically won the right to govern the country in the elections of 1989. Like other foreign spokesmen, Allen was told that Burma would not accept 'outside interference.' Even as the abortive meeting took place, UN officials confirmed that more than 5,000 Karens had fled into Thailand across Burma's eastern border. At this time the KNU made a 'global appeal' for help, saying that it was now sheltering more than 40,000 Karens in Burma who had fled their homes. The Burmese army appeared to have orders to end the Karens' fight for independence once and for all. Units were burning entire villages and killing and raping. The army had press-ganged more than 10,000 Karens including 3,000 women, to become army porters. They were carrying weapons and supplies as the army pushed further into the jungles.

9 February 1992 In an offensive code-named Operation DRAGON, up to 20,000 troops laid siege to a key jungle position on a fortified mountain named Sleeping Dog Hill, eight miles from Manerplaw. It was held by probably no more than 100 guerrillas, while 4,000 others were harassing the Burmese troops. The government appeared more determined to wipe out all opposition than at any time since 1988. The Karens were reinforced in this crisis by 500 troops of the All Burma Students Democratic Front (ABSD) but the students suffered many casualties on Sleeping Dog Hill from howitzers, rocket launchers, Carl Gustav shoulder-held rockets and mortar fire. If the Burmese troops captured the hill they would have Manerplaw itself within range, but the soldiers would then have to cross the Salween river to capture the town.

According to Karen sources, Chinese officials were advising the Burmese at artillery bases near Manerplaw. In mid-March, Sleeping Dog Hill fell to the overwhelmingly strong Burmese forces and Manerplaw and the Karens' frontline trenches came under artillery fire.

10 February 1992 Two boats ferrying 160 Muslim refugees sank in the Naf River after Burmese soldiers fired on the vessels. Few of the refugees survived.

February–March 1992 The Naga Offensive. Until the end of 1991 the Burmese leaders in Rangoon had largely left the 500–1,000 Naga guerrillas alone in their difficult terrain in the north. From here, the Nagas attacked security forces in Indian Nagaland. In December 1991, two Burmese columns, supported by helicopter gunships, had pressed up the mountains, quickly capturing two Naga camps. They fired on the fortress of Longkai with rockets and mortars and on 31 January launched a final assault on Chuiyang Nokn, the guerrillas' political HQ. The offensive was aimed at clearing the way for a much stronger Kachin army in the north. Indirectly, this was a move to capture the gem and timber countryside.

The Rangoon regime's co-ordinated violence on many fronts might appear to be sound strategy, but according to sources in Rangoon it is partly designed to satisfy the ambitions of ambitious regional commanders and to divert attention away from Burma's economic ruin.

The War in 1994

Resistance to Burma's military junta by the various ethnic groups and the students diminished late in 1993 and by 1994 was falling apart. This was the result of a two-pronged strategy by the junta: promises of autonomy to the minorities, and threats of overwhelming and bloody confrontation if these minorities did not meet the junta's wishes.

Despite knowing that troops might be sent against them, the major ethnic groups are edging towards negotiations. The Kachins signed a ceasefire with Rangoon in September 1993 and this weakened the resolve of the Mon, Karen and Karenni resistance groups. They began to talk, at least among themselves, of 'compromising' with the junta.

In January, Karen spokesman Dr. Em Matha said that the Opposition co-ordinating group, the Democratic Alliance of Burma, had accepted that the talks between the SLORC and the Karen had to take place. He said: 'It is very difficult for the Karen not to enter peace talks with the SLORC since all the other ethnic groups are already talking to them'.

The SLORC's military strength has never been in question but foreign diplomats in Rangoon now see the junta as 'much more secure' while its policy to undermine the ethnic insurgents is 'very successful'.

Threats by the Thai authorities to push back across the border about 60,000 displaced Karens and 15,000 Mons have added to the pressure on the Karen leadership. The Thai move came as shock to them; for many years Thailand has allowed the ethnic groups to exist along its border with Burma.

In January the All-Burma Students Democratic Front called for a united resistance front, pledging to continue its struggle until the emergence of a democratic federal union in Burma. However, the SLORC has created divisions among the students.

In December 1993 a report from the London-based International Institute for Strategic Studies argued that on humanitarian grounds and in defence of Western political values a better case could be made for UN intervention in Burma than in Bosnia-Herzegovina. Intervention in Burma would have a clear aim, a finite end and the support of the majority of the people.

'It would right a wrong but will not be undertaken', the report stated, 'despite the illegitimate government and vicious oppression of the people, because the media has not focused the attention of the citizens of the Western democracies on these injustices. Because those governments are not under pressure and because they cannot see their security interests under threat their moral outrage is muted.'

Increased Strength of General Khun Sa

General Khun Sa, responsible for flooding the West with drugs from South-East Asia's Golden Triangle, has become so powerful that he is now a law unto himself in his guerrilla region. The Thais have put a price on his head and in the United States the indictment against him as a drugs dealer grows longer. In a show of defiance at New Year 1994, Khun Sa invited several Thai intelligence officers to his lavish party. The festivities coincided with his declaration of independence from Burma by Shan State's 'restoration council'. Not surprisingly, the council elected Khun Sa as the state president.

Undeniably, Khun Sa has delivered the Shan minorities from persecution and slavery by Burma. He has built schools and established a textile factory and a gem-cutting factory in his zone of control around Ho Mung. Finance comes from the heroin trade. Early in 1994 Khun Sa made yet another offer to the US to destroy the opium crop in return for hundreds of millions of dollars. Later, in a letter to President Clinton, he proposed an end to poppy cultivation if the US and the international community would help his people switch to an alternative livelihood.

Washington's view is that Khun Sa is a ruthless and wily criminal responsible for spreading misery among millions of drug addicts, despite his success at selling himself as a patriotic military leader. The capture of the former Panamanian dictator General Manuel Noriega and the death of Pablo Escobar, the leader of the Colombian drug cartel, suggests that Khun Sa might meet a violent end, despite his bodyguards and army.

References

1. The Diplomatic Corps in Rangoon greatly admires the courage of Aung San Suu Kyi. Envoys say that the harassment to which she is subjected would break most people. 'Her quiet dignity and her resolve to stay in Burma until she can lead her people out of their fearful plight drives the unlawful rulers to fury', one diplomat said. Aung San Suu Kyi owes much to the Diplomatic Corps. It is their watchfulness and outspoken support which protects her; without it she would be murdered by the SLORC.
2. This forcible and sometimes brutal resettlement has gone largely unnoticed in the West and few journalists have reported it. People subject to transfer are given no warning; the first they know is when army trucks arrive to carry out the move.
3. This is the opinion of R.J. Gundala, an Indian journalist with experience of similar movements in Cambodia, Angola, Sri Lanka and Kurdistan.
4. The British Foreign and Commonwealth Office, London, issued a background briefing paper, *Burma: A Missed Opportunity*, in July 1991. Although carrying the disclaimer that the paper 'is not and should not be construed or quoted as an expression of government policy', the Burma briefing is strongly worded and quotes the British government as calling for 'an end to the shameful state of affairs' in Burma. This is certainly an indication of policy.
5. The Shan official was speaking to Reuters' correspondent, who filed a despatch on 10 September 1991.

Cambodia's On-Off-On War

ENTER THE UNITED NATIONS
Background Summary

The Vietnamese community of the old 'Indo-China' of South-East Asia saw itself as the natural successor to control this great region following the defeat of the French at Dien Bien Phu in 1954 and France's consequent withdrawal. The Americans attempted to keep South Vietnam, Laos and Cambodia out of the Hanoi Communists' control, only to be politically and militarily humiliated in what we now call the Vietnam War. Following the defeat of the US in 1975, the unified Vietnam (the North took over the South) had no difficulty in dominating Laos. In Cambodia, the barbarous Khmer Rouge Communists had beaten the US-supported government and as a result Vietnam sent its armies into Cambodia, in December 1978, 'to protect the south-western flank from foreign dominance'. This was a thinly-veiled reference to China, Vietnam's traditional enemy.

Vietnam was allied to the Soviet Union, the enemy of China. China took as an ally the ultra-Maoist Khmer Rouge which, under the infamous Pol Pot, had killed two million Cambodians out of a population of seven million. The victims were people who opposed Pol Pot, those who were suspected of opposing him and, in the case of children, those who might at some time in the future oppose him.

President Heng Samrin put Cambodia's army under Vietnamese command and the opposition groups took to the jungle. They comprised the Khmer Rouge, Son Sann's Khmer People's Liberation Front (KPLNF) and Prince Sihanouk's *Armée National Sihanoukienne* (ANS), both of which are non-Communist.

Under Vietnamese military pressure, the Khmer Rouge lost its mountain stronghold and the KPLNF and ANS lost all their bases. Retreating into Thailand, the resistance groups established bases there as well as camps for their 300,000 refugees. The Vietnamese then vainly tried to seal off the 450-mile Cambodia–Thailand border against guerrilla incursion. In the severe fighting that followed the Vietnamese army lost 25,000 soldiers. Under pressure from the Soviet Union, Vietnam withdrew its troops in 1990.

The Khmer Rouge quickly took advantage of the Vietnamese withdrawal, formed a 'coalition' with the other resistance groups, and began what the Cambodia government of Prime Minister Hun Sen called 'the war of the villages'. The Khmer Rouge assumed the highly inappropriate title of Party of Democratic Cambodia and claimed that its revolutionary war was at an end. It was now engaged in a 'people's war' to liberate the country.

Pol Pot was supposed to have retired and to have been succeeded by Khieu Samphan. However, the most powerful figure was Son Sen, vice president of the movement and its army commander. The US government was caught in a dilemma, largely of its own making. It wanted to support the anti-Communist groups but feared that any arms sent to them might finish up in the hands of the Khmer Rouge.

During 1990 the character of the war changed. Because the Vietnamese troops had gone, combat was no longer confined to sporadic fighting in the region of the Thai border. Moving far out from their bases on the Thai side of the border, the guerrillas pressed towards the capital, Phnom Penh. The threat from the Khmer Rouge was so great that the government resettled the entire population of many villages in places where the army could more easily protect them.

As leading countries tried to promote a peace settlement the resistance groups began an offensive and in June 1990 captured areas of several provinces, including Kampong Speu, Siem Riep, Battambang and of the strategically important Kampong Thom. Khmer Rouge atrocities became commonplace and the US withdrew its support for the resistance coalition. This was a dramatic enough step but the Bush administration went further: it announced that it would reverse a decision made during the Carter presidency and open talks with Vietnam about Cambodia. Secretary of State James Baker explained that the Bush administration's strategy was to secure the total withdrawal of Vietnamese forces from Cambodia (this came about in September 1990), to prevent the Khmer Rouge's return to power and to bring about free elections for a new government in Phnom Penh.

The US was supported by the Soviet Union but the ASEAN countries — Singapore, Malaysia, Brunei, Indonesia and the Philippines — denounced the plan. ASEAN stated that recognition of the anti-Communist anti-Hun Sen coalition was dangerous and would set back the cause of peace.

The War in 1991–92

The pattern for what would happen in 1991 was established in September 1990 in the southern Chinese city of Chengdu. Here China, which supports the Khmer Rouge, and Vietnam, the main backer of Phnom Penh, held a meeting to try to bring about peace. In fact, the meeting succeeded only in hardening the already rigid attitudes of both sides.

As a result, the Chinese increased their aid to the Khmer Rouge and encouraged the movement to commence a new advance. In the first two months of 1991 Vietnam deployed combat units in Cambodia to block the Khmer Rouge. Meanwhile other conflict took place, indicating the growing tensions between groups in Cambodia. For instance, in Battambang on 10 February 16 civilians were killed during fighting between government troops and the city's police force. This affair was not as unexpected as it might appear. The grossly underpaid army conscripts resent the prosperous merchants, who pay the police well for protection. Soldiers not infrequently extort jewellery from market traders at gunpoint and this causes fighting between police and soldiers. In one incident in Battambang about 20 shells fell on the market occupied by the gold merchants and caused many casualties. The army put about the story that the Khmer Rouge was responsible and the Khmer Rouge itself was happy to be blamed because their involvement would terrify the populace.

Discipline in the army declined steadily during 1991. The peasants conscripted into the army and forced to bear the brunt of war have always hated the rich and the relatively rich of the towns, who somehow manage to evade military service. The indiscipline and low morale of the soldiers led the Ministry of Defence in Phnom Penh to bring in large numbers of Vietnamese army veterans to defend some towns from the insidious advances of the Khmer Rouge, notably Battambang and Siem Riep. Throughout these activities the government denied that Vietnamese troops were in the country but government officials privately admitted that the Vietnamese were back. The government was paying the equivalent of $60 a month each to the foreign troops on 'passive service' but when they were engaged in active service the payment was increased to $250 a month.

On 1 May, the UN Secretary-General, Javier Perez de Cuellar, made a personal appeal for hostilities to stop and, rather to international surprise, his request was assented to. Within 10 days a four-man UN team arrived to inspect the activities of all the warring factions. Its leader, Major General Timothy Dibuama, from Ghana, met Khmer Rouge officials at a secret camp in Cambodia near its border with Thailand. On 13 May he visited bases of the non-Communist guerrilla factions and later he inspected government troops. The team said that their mission was largely symbolic and this seemed to be the impression of the Khmer Rouge, which sent only junior officials to the meeting. The mission's aim was to show the Cambodians that the UN was anxious to implement a full-scale peace plan.[1]

While the UN mission was in the country, Khmer Rouge radio repeated familiar charges that the Phnom Penh government was allowing millions of Vietnamese to settle in Cambodia as part of a strategy to 'swallow' the country. Virtually at the same time Phnom Penh radio accused the guerrillas of violating the ceasefire and threatened to 'punish them in order to defend the people's lives'.

The ceasefire lasted a month, one of the longest periods of relative peace throughout the 12-year conflict. It was broken by government troops launching an attempt to capture the Khmer Rouge-controlled town of Pailin, near the border with Thailand, before the summer rains brought an end to fighting.

Hun Sen and his government wanted to do more than capture a town. Pailin is the centre of the ruby-mining industry, and using the town as a base, the Khmer Rouge had for several years been selling mining concessions and making much money from them.

The Pailin operation, as well as territory-grabbing movements by the non-Communist resistance factions, ended talks in progress in Jakarta, Indonesia, aimed at finding a peace formula. They had not been going well in any case. Hun Sen's delegation insisted that Khmer Rouge leaders should be tried for genocide and that the movement should be barred from the elections contemplated in the UN-sponsored plan. The Khmer Rouge refused to endorse an extension of the ceasefire. Diplomats said that the movement's isolation was

Cambodia: Khmer Rouge Activity Continues Despite Peace Efforts

significant. Just how significant is in doubt. The movement's military strength made it more than capable of continuing the civil war.

Soldiers Became Bandits

Cambodia is in a desperate situation. There was a $50 million deficit on a $150 million budget in 1991 and the government relied on the Sino–Khmer merchants to keep the economy afloat. They import oil products and fertilisers, for instance, in exchange for concessions to export rubber and other raw materials in deals that yield profits up to 400 per cent. However, no wealth reaches the people and the annual *per capita* income is only $110. Economic deprivation produces more casualties than the war itself. Jean-Jacques Fresard, chief representative of the International Red Cross in Phnom Penh, reported that child mortality was the highest in the world, with 20 per cent dying before the age of five. The rate of tuberculosis was double that of the rest of the Third World.[2]

Diminishing aid from the Soviet Union has made the situation even worse. In 1991 Cambodia had not built a single new structure for its citizens in 12 years; schools had no books, clinics had no medicine. Many doctors, teachers and other civil servants were waiting months for their $7.00 a month salaries.

Travellers reported that peasants, small shopkeepers and even government officials were talking wistfully of 'the good old times', meaning the comparatively prosperous and peaceful period before Prince Sihanouk was overthrown by a military coup in 1970. Now, Sihanouk can offer no solution while he is in alliance with the Khmer Rouge.

In August 1991, at the Thai resort of Pattaya, the factions agreed on a partial demobilisation of forces. This crucial decision was reached only after a compromise that allowed each group to retain 30 per cent of its troops. This raised fears of lingering security problems.

It was soon clear that demobilisation was not the solution which so many people had expected. The abandoned or discharged troops turned into freebooters and lawless armed gangs. They took to the countryside looking for anything they could get their hands on. Murder has been commonplace as bandits equipped with automatic weapons terrorise villages and rob road convoys of emergency supplies trucked in by the relief agencies.

Troops loyal to the Phnom Penh regime deserted in droves as funds from Vietnam and the Soviet Union dried up. At times the marauders seized control of the vital north–south route to the port of Kampong Som, which supplies most of Cambodia's meagre trade. Wandering groups of former soldiers ransacked shops and houses up to 60 miles inside Thai territory. The Thai army tried vainly to plug the many trails which cross the border.

The Khmer Rouge has not suffered the same desertion rate because it has been able to bind its force together with promises of incentives. They offered land, jobs and even cash to keep the fighters together and announced plans to develop new settlements where its people could be kept together.[3]

The Khmer Rouge accused the government of filling its dwindling army ranks with children in its desperate attempts to maintain its troop strength. Aid workers reported that all four factions were drafting women and children to replace defecting troops.

UNTAC and its Mission

Despite all the preliminary difficulties, the UN plan came into being, known as the United Nations Transitional Authority in Cambodia (UNTAC). In October 1991 the mission was expected to cost up to $5 billion over two years and require 30,000 soldiers, civilian administrators and relief workers. After the settlement was signed in Paris, on 31 October, UNTAC moved into Cambodia to supervise the post-war transition. A peacekeeping contingent of 20,000 troops drawn from several countries, including Australia, Austria, Canada and Sweden, supervised the disarming and partial demobilisation of the factions. About 70 per cent of the former fighters will return to civilian life. The remaining 30 per cent will be garrisoned, with their weapons, in UN-guarded and supervised cantonments.

About 350,000 of the country's 8.3 million people are armed, though less than half that number are thought to belong to organised units. Disentangling guerrilla armies and determining whether people in civilian clothes are really fighters are tasks no previous UN force in other war-torn countries has tackled. The Khmer Rouge tried to hide as many fighters as it could from UN inspectors.

Civilian UN officials were trying to organise elections in a country that has no census or free press and little democratic tradition. To keep the Hun Sen administration from using its position to influence the election campaign, the peace plan called for UNTAC to put foreign officials in key government jobs until the ballot took place. After the elections the assembly would write a constitution and then turn itself into a legislature.

The UN mission supervised the return of 300,000 Cambodian refugees from camps along the Thai border. Officials had hoped to keep the flow to 10,000 a month but refugees were making their way home as early as September 1991 for fear of losing the best land to those who arrived first. The trek back into Cambodia and resettlement there is dangerous because of the profusion of landmines, perhaps one million of them.

All tasks in Cambodia are complicated by the lack of an infrastructure. Roads and bridges are in rough condition, power supply is erratic, housing is scarce and a telephone service is virtually non-existent.

Whether the strategy would work was doubtful. The Khmer Rouge was dragged into the settlement but has not repudiated the genocide it perpetrated. The leaders who directed it are still in power and they control huge quantities of weapons from China. The Khmer Rouge acquiesced to peace only as a temporary manoeuvre and UNTAC very nearly found itself in a war. It is possible that the Hun Sen government and the Khmer Rouge, provoked by Hanoi and Beijing, will collude to impose a Communist regime on Cambodia.[4]

British Government's Admission of Involvement

On 26 June 1991, the British government admitted for the first time that it gave military training to armed Cambodian resistance forces for six years in the

1980s. A written answer in the House of Commons from Archie Hamilton, the Armed Forces Minister, revealed that training had been halted in 1989. But it confirmed for the first time that Britain had trained the *Armée Nationale Sihoukienne* and the KPNL.

The official answer coincided with the announcement that the Khmer Rouge and the non-Communist factions were to return to Phnom Penh to form a Supreme National Council. Under the terms of an agreement, this Council, which will be chaired by Prince Sihanouk, was intended to represent Cambodia's sovereignty until elections could be held to form a new government.

Mr. Hamilton's statement repeated that 'there has never been and will never be any British assistance or support, military or otherwise, for the Khmer Rouge'. However by confirming that Britain trained the ANS and the KPNL, Hamilton's answer was in contrast to earlier refusals by Foreign Office ministers under the Thatcher administration to be drawn on whether Britain had helped other participants in the Cambodian civil war.

Hamilton said that 'in accordance with normal practice' he could give no further details of training. In fact, it was carried out by the SAS. The minister disclosed that training had started in 1983, earlier than previously thought. Since 1989, he said, 'Her Majesty's Government has not in any way been involved in training, equipping or supplying the forces of any of the Cambodian parties'. The purpose of British training for the non-Communist resistance had been to strengthen the position of those forces in relation to the more powerful forces of the Khmer Rouge in their struggle against the Vietnamese-imposed regime in Phnom Penh.

The Labour Party's defence spokesman, Martin O'Neill, said that the KPNL and the ANS had fought 'side by side with the Khmer Rouge with the common objective of getting control in Cambodia. The government has therefore indirectly helped Pol Pot and the Khmer Rouge in their attempt to regain power. No excuse can justify this inhumane policy.'[5] The Labour Party did not allege that British assistance was being given directly to the Khmer Rouge and the government has consistently denied such claims. The disclosure by Mr. Hamilton renewed controversy over long-held claims by the journalist John Pilger that British special forces had directly assisted the Khmer Rouge.

The War in 1993-94

The elections in May 1993 were a success despite the boycott imposed on them by the Khmer Rouge. On the whole the ballot was fair and democratic, though many of the voters had no conception of the meaning of democracy. The UNTAC mission was a personal triumph for Lieutenant General John Sanderson, the Australian in command. With great courage and skill, he led his military force of officers and men from 28 different countries through the political minefields that confronted them and won respect even from the Khmer Rouge. Sanderson's success once again demonstrated the need for a carefully chosen senior leader for any UN mission. This might seem obvious but many leaders have been abject failures. With his clever blend of firmness and compromise, Sanderson emerged from his Cambodian mission with an international reputation.

The Khmer Rouge continued to be a threat to Cambodia's democracy and stability. Early in January a senior government officer said that army forces would capture the Khmer Rouge's northern HQ during the month. Phnom Penh radio announced: 'The headquarters of the Royal Cambodian Armed Forces appeals to all DK [Democratic Kampuchea or Khmer Rouge] officers and rank-and-file urgently to return to the national community in order to avoid imminent confrontation.' Cambodia's two prime ministers, Prince Norodom Ranaridhh and Hun Sen, appealed for national reconciliation.

Indeed, more than 3,000 Khmer Rouge had defected to the government side since August 1993, bringing with them many weapons. Phnom Penh radio made much of this, though it exaggerated the numbers of men and their weapons.

The man responsible for planning the attack on the Khmer Rouge was General Long Sopheap, GOC Fourth Military Region, which includes the strategic frontline provinces of Siem Reap, Preah Vihear and Kompong Thom. His main objective was to take the Khmer Rouge's base at Anlong Veng. Broadcasting on Phnom Penh State radio, the General said that there had already been clashes between the army and guerrillas in Chikreng and Siem Reap districts. Other Cambodian sources said that the preparation for the government assault was the biggest in five years. It was also the first to be supervised by the new coalition government, led by the royalists.

Two days after General Sopheap's warning, on 6 January, the army moved. About 1,200 troops supported by tanks advanced towards Anlong Veng, 130 miles north-west of the capital. No Western observers were able to accompany this thrust but those based on Phnom Penh doubted the ability of the army to reach Anlong Veng without a much greater force.

They were correct in this judgement but they misjudged the government's resolve. The Khmer Rouge repulsed the army's initial advance, whereupon the government did what it should have done in the first place. It assembled 7,000 troops, together with tanks, helicopters and multiple rocket launchers. The army this time chose a different objective, the Khmer Rouge base at Pailin, near the Thai border. By the end of the month the army had captured enemy positions in the surrounding hills and from there moved against the defences in Pailin and captured them.

Western diplomats said that taking Pailin could be a mistake for the government because it would be difficult to hold. It was obvious that the Khmer Rouge would exert every effort to retake the place because its ruby mines and timber stocks bring the organisation millions of dollars. Having captured it from the Vietnamese-installed government in 1989 they made it the power base for their insurgency.

There was every sign, in mid-1994, that despite some government successes the Khmer Rouge remained too strong to be crushed. As always, the strength cannot be assessed only by the number of men under arms; much of its power continues to come from its ability to terrify the ordinary Cambodian.

References

1. General Dibuama proved himself to be a successful diplomat. He gained greater trust from the always suspicious Khmer Rouge leaders than any negotiator in recent years. Dibuama

had the advantage of being 'untainted', in Khmer Rouge eyes, by association with Western interests.
2. M. Fresard was speaking to the author. The IRC unofficially regards Cambodia as its greatest war-related problem.
3. Pol Pot's influence is seen in every action taken by the Khmer Rouge. Though one of the most evil men of the century, he is paternalistic towards his own followers. This paternalism is the Khmer Rouge's binding force. The movement's propaganda often refers to 'the great Khmer Rouge family and its generous father'. The 'father' is the butcher Pol Pot.
4. This suspicion was widespread in the Diplomatic Corps in Phnom Penh at the end of 1991. One diplomat said: 'it is impossible to believe that a democracy can survive in Cambodia for longer than a few months.'
5. Details from *Hansard*.

Aspects of the War covered in War Annual No. 5 include:
Khmer Rouge atrocities.
Analysis of the Cambodian Army.
Vietnam, the peace process and ASEAN.
The 'Christian Coup' pot.

The Central American Arena

OVERVIEW

Costa Rica, El Salvador, Guatemala, Honduras and Nicaragua — the countries of Central America — have been afflicted by war since the 1950s. Conflict has not necessarily affected them all equally or at the same time but the various armed struggles have impeded the development of the entire region and have seriously damaged the social and economic structure.

The United States and the Soviet Union — when it was still a superpower — have been deeply involved in support of either a client government or of their opposition. The US supports non-Communist states for no other reason than they *are* non-Communist. But some of them, notably El Salvador and Guatemala, have been guilty of flagrant breaches of human rights and the murder of tens of thousands of their own citizens. Washington's closest ally, Honduras, in 1988 became the first country to stand trial before the Inter-American Court on Human Rights.

In January 1987 President Arias of Costa Rica put forward proposals to bring about peace in Central America. The 'Arias Accord' became the Guatemala Peace Agreement on 7 August that year. Its most significant clause reads:

> The governments of the five Central American states request governments from outside the region which are providing either covert or overt military, logistical, financial or propaganda support, in the form of men, weapons, munitions and equipment to irregular forces or insurrectionist movements, to terminate such aid. This is vital if a stable and lasting peace is to be attained in the region.

This was a plea more than a request and it was aimed mainly at the US, the Soviet Union, Cuba and Spain. Without their interference and aid it is doubtful if any of the wars could have lasted as long as they did. Within the unfortunate Central American countries, entrenched authority and privilege was in conflict with the aspirations of the poverty-stricken majority. For everybody it was necessary to fight a war to survive. Compromise was an impossible concept.

The conflicts in Central America are basically about power, whatever their precipitating causes might have been. This has also been the case with the superpowers, each of which was alarmed by the political power it perceived — and, therefore, the military power as well — the other was developing. The Reagan administration's policies were extreme; no cost was too great to stop the spread of Communism. The Soviet Union had a different problem: its

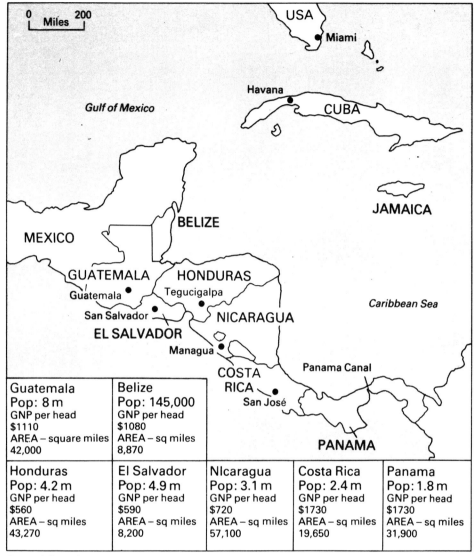

Central America

clients expected so much from it that it was forced to go to dangerous lengths to meet their expectations.

Central America was not important in world politics but the leaders of the superpowers treated it as though it were. President Reagan held the bizarre belief that tiny Nicaragua was somehow so sinister and powerful under the Sandinista government that it might invade the United States. This was nonsense.

Nevertheless, largely because of the UN's involvement in Central America, peace moves were initiated and developed. Within a period of 12 months, four governing parties in Central America lost office to the Opposition through democratic elective processes. It seemed, at the end of 1990, that the region might find itself at peace within a decade.

By 1994 there was still much instability, though American commentators, from both government and media, continued to find the prospects for Central America 'encouraging' or 'improving' or 'no longer troublesome'. These assessments may have resulted from a vision fogged by great upheavals in other parts of the world. For instance, US preoccupation with Somalia, South Africa, the Balkans and South-Central Asia led to Washington's focus on Central America being less sharp than it was up to 1990. President Bush and later President Clinton were forced to relegate Central America to the 'no longer urgent' category.

In fact, El Salvador, Nicaragua and Guatemala have all seen low-level insurgency in 1993 and 1994. The major reason armed conflict has not been more constant and violent is that the US and Russia can no longer afford to intervene in the Central American countries' affairs by supplying them with costly armaments. Former Soviet-satellite countries, such as Cuba, can even less afford to supply arms and equipment, while one Soviet satellite country formerly deeply involved in weapons supply, East Germany, no longer exists.

El Salvador Civil War

TOWARDS A PEACEFUL END

Background Summary

The conflict in El Salvador began in 1980 and grew out of a dispute between the Rightist government of the Christian Democratic Party and several disparate Communist and Socialist movements. These organisations made little headway until Fidel Castro of Cuba induced them to amalgamate. They did this in three unions and as they developed several notable leaders were thrown up.

The chief guerrilla leader of the People's Revolutionary Army (ERP) was Joaquin Villalobos. Other important men were Shafik Jorge Handal, General Secretary of the Salvadoran Communist Party; Eduardo Cantaneda (alias Ferman Cienfuegos) leader of the Armed Forces of National Resistance (FARN) and Robert Roca of the Central American Workers' Revolutionary Party.

President José Napoleon Duarte tried to crush all opposition by the most brutal methods. His death squads killed 70,000 people up to 1988. With much American help, the Salvadoran army became a well-trained and ruthless counter-revolutionary force of 57,000 troops. Even so, they could not destroy the 8,000 guerrillas, whose main assistance came from Cuba and Nicaragua.

In 1988, the ultra-Rightist Republican National Alliance (ARENA) came to power, with Alfredo Cristiani as President. The party's chairman, Armando Calderon, had equal if unofficial power but ARENA's founder, Roberto D'Aubuisson, dominated both of them. D'Aubuisson directed the death squads.

During 1989–90 the army pursued a hearts-and-minds campaign to woo the peasants away from their support of the guerrillas. The rebel leaders themselves showed their contempt for the army by sending 3,500 guerrillas into the capital, San Salvador, in a massive raid. During this raid and the death squad reprisals which followed more than 2,000 people were killed.

Despite the atrocities — and partly because of them — peace talks took place, only to break down on the issue of military reform. The rebel chiefs argued that only great pressure on the armed forces to be less brutal would bring progress. The government insisted that a mutual ceasefire had to precede any political accords. The guerrillas responded by refusing to lay down their arms until the government agreed to dissolve the military over a stated period, replace the hated security forces with civilian police and punish human rights violators within the army.

How the War Ended

It seemed in June 1990 that the deadlock in negotiations could never be

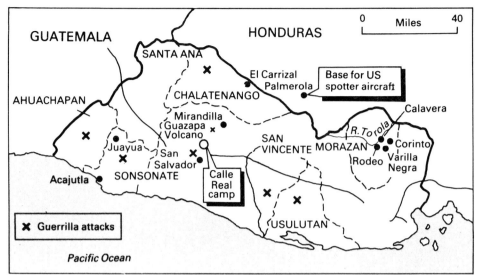

El Salvador Civil War

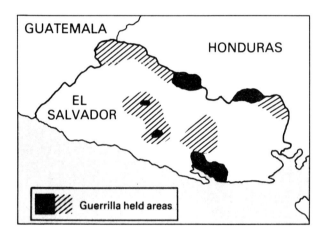

broken. Each side believed that if it held out long enough it could win the war and each so hated and distrusted the other that no progress was possible.

The first event outside the country affecting the war inside was the defeat of the Sandinistas of Nicaragua, the main suppliers and supporters of the Salvadoran guerrillas. Having lost power in their own country, the Sandinistas were suddenly in no position to help their friends beyond the border. This was a devastating blow to Villalobos and his friends.

But worse was to come. The break-up of Communist Eastern Europe and the collapse of Communism in the Soviet Union itself presaged a decision in Moscow that would not have been credible a few years earlier. Moscow withdrew most of the aid that had gone to Fidel Castro of Cuba. Castro was the Soviet Union's conduit, the surrogate paymaster for Central and Latin America. In addition, whenever Castro had received modern military equipment from the Soviet Union he had offloaded the outdated weaponry onto the FMLN. Now, having been warned not to expect any further shipments, Castro needed all his weapons and equipment.

Mr. Perez de Cuellar saw his opportunity. Until now peace talks had been through third parties to the El Salvador dispute — Mexico, Spain, Costa Rica, Venezuela and the UN. The UN Secretary-General was now able to induce El Salvador's President Cristiani and the leaders of the co-ordinating organisational body, Farabundo Marti National Liberation Front (FMLN) to talk face to face.

They agreed on these key points:

- Representatives from the Salvadoran government and the FMLN would determine the criteria for reducing the armed forces.
- An agenda would be set up for the creation of a national civil police force and the dismantling of the security forces.
- The mission of the armed forces would be redefined, focusing on the defence of the nation's sovereignty, protecting the integrity of its territory and respecting its human rights.
- A national peace commission would be created to supervise all political agreements and relations with the armed forces. Its membership would comprise two rebel representatives and one from each party in the National Assembly.

While these developments were under way the Bush administration quietly told the Salvadoran government that military assistance would be greatly reduced. A senior State Department official was sent to San Salvador to tell Cristiani and his even more belligerent colleagues that the US could no longer encourage their efforts to suppress political opposition by force. As in Eastern Europe, so in Central America democratic institutions must prevail.

'The Gordian knot has been untied', said Perez de Cuellar at the time of the initial signing of the agreement on 2 October 1991. A formal ceasefire agreement with attention to many fine points had then still to be drafted but the vicious war was apparently in its dying phase.

The Conflict 1992-94

The United Nations sponsored a peace agreement in 1992 and the ARENA government and FMLN revolutionary movement agreed to democratic elections in 1994, the first in El Salvador this century. The election campaign was so violent that Jeremy Hobbs, executive director of the Australian Development Agency, Community Aid Abroad, suggested that 'the elections are only the initial step in a longer road to democracy'.

The government's death squads were in action again and more than 30 political murders occurred during the campaign. Nearly all of them were of opposition candidates. The death squads were openly linked with senior military officers and the government.

US State Department documents declassified in 1993 link the government's presidential candidate for the election, Fernando Armando Calderon Sol, to death squads in the 1980s. About the same time as the Calderon disclosures, the UN Truth Commission found that the ARENA party founder and death-squad leader, Roberto D'Aubuisson, had directed the murder of El Salvador's Archbishop Oscar Romero in March 1980. D'Aubuisson's involvement had long been suspected.

Two people were killed in shootings during an opposition rally. A government campaign march degenerated into chaos when street vendors pelted the government entourage with tomatoes and fruit. Elsewhere local youth gangs assaulted government supporters, resulting in 50 casualties. UN observers called all this 'war on the streets'.

The body responsible for the processing of electoral registrations - a process that easily leads to corrupt practices - is the Supreme Electoral Tribunal, a highly politicised organisation. It is likely that because of bureaucratic ineptitude, together with vote-rigging, more than 400,000 dead people remain on the electoral rolls. The Electoral Tribunal announced that no polling booths would be placed in the Opposition stronghold of rural Chalatenango. This meant that the Opposition would lose vital support as many voters would be unable to make the long trip to vote in the provincial capital. Head of the UN operation in El Salvador, Augustus Romero Campo, denounced the decision as 'violation of the electoral code'.

The government also caused controversy by planning to call out the military on election day - 'to provide security'. The air force transported the ballot boxes, giving the government another chance to tamper with the votes.

Experienced observers in El Salvador said that no matter which party won the election the seeds of a continuing civil war had been sown. The losing party would not accept the electorate's decision as final. Augustus Romero Campo said: 'Trust does not exist in El Salvador. Without it the conflict will go on.'

In the elections, held on 22 March 1994, ARENA's candidate, Fernando Calderon, needed more than 50 per cent of the votes; he polled 49.5 per cent, leaving Ruben Zamora, presidential candidate of the Left-wing coalition, the victor. However, the Left's chances of a victory on the second ballot on 24 April seemed remote. Smaller parties, including the Christian Democrats, which won about 15 per cent of the general vote, were more likely to make an alliance with ARENA. In any case, fraud would deprive the Left of many votes. As was to be expected, ARENA, with the support of the smaller parties, did emerge victorious in the second ballot, which was riddled with irregularities.

Nevertheless, the FMLN's commander, Joaquin Villalobos, was satisfied enough and spoke with confidence: 'We have consolidated our position as the second political force in the country,' he said. Whether the left can hold together following the election is doubtful. Apart from the strains within the coalition, the FMLN is itself a front, made up of five different organisations.

Guerrilla War In Guatemala

PROFESSIONAL REBELS

Background Summary

Since 1954 Guatemala has known nothing but war and civil violence. Throughout this period Left-wing guerrilla groups have fought the ultra-Rightist government for basic land reform and wealth redistribution. In response, the government has encouraged the security forces to massacre civilians suspected of providing the guerrillas with food and shelters.

The four guerrilla groups, after many years of mistrust and antagonism, amalgamated in 1985 to become the Guatemalan National Revolutionary United (GNRU). Even so, the Organisation of the People in Arms (ORPA), led by Rodrigo Asturias, remained the most dominant. Alarmed by guerrilla successes, the government in 1986 turned virtually every village into a fortified army post. All able-bodied men were forced into 'civil defence patrols'.

The Council of Ethnic Communities (known by the Spanish acronym of CERJ)[1] advised the peasants to exercise their constitutional rights and not to serve in these patrols. The security forces harassed CERJ's leader, Amilcar Mendez Uruzar, and murdered other CERJ officials.

A former president, General Meija Victores, urged President Vinicio Cerezo Arevalo not to be 'foolishly humanitarian' in his handling of Guatemala's problems. Victores, rather than Arevalo, was supported by Colonel Luis Arturo Isaacs Rodriguez, the official military spokesman.

The War in 1991–92

While conflict fell away sharply in Nicaragua and El Salvador, it remained at a high level in Guatemala and the amount of actual fighting and the number of casualties may have increased. One important fact about guerrilla warfare became clear — that large numbers of guerrillas are not essential to hold down large numbers of troops. This certainly applies in Guatemala itself, where the many guerrillas are veterans and where army intimidation has not cowed the people as a whole. There may be no more than 2,000 guerrillas, fewer than in 1989, yet they continue to show remarkable resilience and professional military ability.

During 1991, in addition to the long-standing effort to crush the Indians, who have always been accused of 'collaborating' with the guerrillas, the army embarked on yet another campaign of violence. This campaign of intimidation, harassment and sheer terrorism was aimed at small but vocal and influential groups accused of supporting the guerrillas, including trade union and student leaders and human rights advocates. The man behind the terrorism was

Colonel Rodiguez. President Cerezo is known to oppose such methods but he is a weak man and as a civilian he has no support from any senior army commander.

The President is so malleable in the hands of the army and police that he personally intervened on their behalf when they were jointly under investigation for involvement in narcotics trafficking. Despite strong evidence from witnesses of exceptional integrity, Cerezo had the court case stopped.

Guatemala's guerrillas have not been as dependent as El Salvador's rebels on help from the Soviet Union, Cuba and the former Communist countries of Eastern Europe. Arms and money have come from supporters in South America and even in the US. Also, the GNRU troops have been remarkably successful in stealing supplies from army and navy depots. In any case, a guerrilla 'force' of 2,000 men and women does not need large quantities of supplies when it draws its basic sustenance from the civilian population.

US military support for the Guatemala government remains undiminished and the entire air force fleet is American-built. It includes 10 Bell attack helicopters and 18 Cessna counter-insurgency aircraft. In a secret operation, US technicians and mechanics serviced the entire fleet, which had become run-down, in mid-1991. Virtually all the army's tanks, armoured personnel carriers, towed artillery and mortars are US-built.[2]

The armed forces amounted to 45,000 in 1991, with 42,500 of them in the army. Since service is by conscription, with 30 months service, very large numbers of Guatemalans have had military training. For the government, army service is as much about political indoctrination as fighting the guerrillas. 'Hate sessions' — with the GNRU and its supporters as the target — are an important part of 'military' training.

Guatemala covers an area of about 109,000 square miles. According to adventurous Latin American journalists and foreign human rights activists who made contact with the rebels in 1991, they are in good heart and health and natural wastage due to age is made up by recruitment of peasant youths. One visitor in 1991 described the guerrillas as 'invincible'.[3]

Government and army communiqués claim successes against the guerrillas but it is recognised in Guatemala City that the security forces suffer many more casualties. Official US documents in the hands of Guatemalan-support groups in New York appear to show that American army advisers in Guatemala are frustrated by the army's continual inability to flush out the rebels and by its brutal tactics against the Indian population.[4]

References

1. CERJ remains the best and most reliable source of information out of Guatemala. The organisation runs a 'courier service' to get news out, since the security forces tap telephones at will and mail is frequently opened. The press is censored and under government supervision and no media report should be taken at face value.
2. Guatemala has a very large paramilitary force. The Territorial Militia has an estimated 600,000 members, the National Police 11,000 and the Treasury Police 2,250. The Treasury Police exist to ensure payment of taxes and duties and the force is supposed to fight corruption in the government and bureaucracy. However, diplomats report that the Treasury Police as a body is riddled with corruption.
3. This traveller has a professional military background and his assessment must be taken seriously.

Guerrilla War in Guatemala

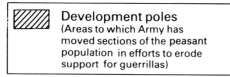

Development poles
(Areas to which Army has moved sections of the peasant population in efforts to erode support for guerrillas)

4. Some American advisers have asked to be relieved of duties in Guatemala. The US government plays down the presence of American personnel and would undoubtedly like to end its involvement in the country but it feels that American withdrawal could lead to other parties moving in.

Nicaragua — The Contra War

CONFLICT IN ITS DEATH THROES

Background Summary

The corrupt Somoza family ruled Nicaragua for decades until the Sandinistas forced the dictator Anastasio Somoza to flee the country in 1979. By then the founder of the rebel movement, Augusto Sandino, had been dead for 46 years. The Sandinista National Liberation Front (SLNF), led by the brothers Daniel and Humberto Ortega and Thomas Borge, gained control. Opposing them were the Right-wing Contras, with the Nicaraguan Democratic Front (NDF) the principal party.

The Left-wing Sandinistas and their army were backed by the Soviet Union and Cuba; the Contras were supported by the US, Saudi Arabia, the Gulf states and Portugal. Fighting took place in the mountains, river valleys and coastal swamps.

Late in 1988 the war ground to a halt when the US Congress cut off military aid to the Contras. The Bush administration protested and only reluctantly endorsed the Congressional decision. Large numbers of Contras remained in training into 1989 but many of their leaders found refuge in the US. President Ortega agreed to Nicaraguan elections being held in February 1990, but on 1 November 1989 he ended a 19-month ceasefire and began a new offensive against the troublesome Contras.

The National Opposition Union (UNO), an amalgamation of 14 rightist parties, put up 2,000 candidates for the national elections. Under Mrs. Violeta Chamorro, UNO won these elections.

This result, which surprised foreign observers as much as it did the Sandinista leadership, resulted in a 14-point agreement. A ceasefire went into effect immediately and the Contras agreed to disarm and demobilise totally by 10 June 1990. By the end of May only 1,700 of the remaining 15,000 armed Contras had handed in their weapons; there were still many armed men holding out in the forests and in Honduras.

In December 1990, only eight months after Mrs. Chamorro became President, 70 armed Contras seized a police post and fought against soldiers who attempted to recapture it. The national army consisted entirely of soldiers from the former Sandinista army and many observers predicted that the war, which was officially at an end, would continue.

The 'War' in 1991

Nicaragua continues to suffer from the strains and stresses which result when a genuinely democratic government has to depend for its military security on

NICARAGUA – THE CONTRA WAR

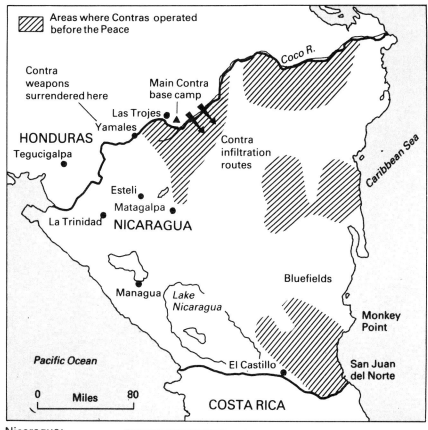

Nicaragua:
Sandinistas
versus Contras

an army which is controlled by its defeated adversaries. The armed forces are known, in fact, as the Sandinista People's Army. This reminds everybody that the Ortega brothers and their supporters could destroy the government whenever they choose to do so.

Throughout 1991 the bitter and vengeful Contras made small-scale attacks against army and police posts but no organised resistance was possible because few leaders remained in Nicaragua and, except for humanitarian assistance, foreign aid had dried up.

Diplomats in Managua are sure that the Sandinistas are biding their time for a return to power, that they merely need an excuse to overthrow the government. A source in the capital says that the Sandinista leadership is trying to find a way to incite the Contras to make more frequent and more damaging attacks on government installations. They would then claim that the government was incapable of governing the country and seize power on the pretext that only they, with the backing of a loyal army, could do so. When this happens, the source says, General Humberto Ortega, as the national army commander, would become the dominant figure in Nicaragua, rather than his elder brother, Daniel, the former President.

However, a new stabilising force is now at work. The Left-wing Sandinistas now have no foreign government to support them with money and arms. The Soviet Union no longer exists and the former Soviet republics, which are now largely democratic and are going their own way, have no interest in spreading Communism in Central America. With this threat removed, the US government accords Central America only a low political priority. The Contra war might genuinely be at an end, or dying, simply because no outside power is stoking it. President Arias of Costa Rica had said that war would end in Central America if foreign powers stayed out of the region. He seems to have been proved correct.

The United Nations monitoring force (ONUCA), which is also deployed in Honduras and El Salvador, is based on a Venezuelan infantry battalion. It is supported by a Canadian helicopter detachment, civil medical units from Germany, four fast patrol boats from Argentina and observers from nine countries.

Colombia Civil War

'THE WAR THAT WILL NOT END'

Background Summary

The conflict in Colombia began in the late 1950s as a struggle between liberals and conservatives. Various armed groups came into being and the war which developed grew increasingly complex. The most important warring factions were:

- Colombian Armed Forces or *Fuerzas Armadas Revolucionarias Colombianes* (FARC), the military wing of the Communist Party, led by Manuel Marulanda.
- 19th April Movement, always known as M-19, the most powerful group until displaced by the Army of National Liberation. Its leader was Carlos Pizaro.
- Army of National Liberation (ELN), a pro-Cuban group. It lost its separate identity after 1986 when it merged with the National Guerrilla Co-ordination (CNG). CNG was led by a Roman Catholic priest, Father Perez.
- People's Liberation Army (EPL), a Maoist group.
- Workers' Self-Defence, known in Colombia as ADO.
- The Patriotic Union, a Right-wing organisation of government officials and supporters.

In 1988–89 another 130 Right-wing paramilitary organisations were functioning.

The narcotics barons and their private armies made the wars of the 1980s even more complex. The most notorious of them, Pablo Escobar, declared 'total war' against the presidency of Virgilio Barco and in 1988 Colombia was in a state of anarchy. Many military commanders became so terrified of the guerrillas that they ceased to be effective.

M-19 and FARC lost so many members that the groups voluntarily disbanded in 1990 but the wars between the drugs barons and the government forces became even more ferocious. Casualties were numerous but the 'Elite Force', created by the Colombian Police commander, General Miguel Antonio Gomez, gained ascendancy over 'the men of Medellin', the drugs chiefs.

In April 1990 Carlos Pizaro was assassinated while on a flight in Colombia. His death had been ordered by the drugs barons, working with fascist groups. Pizaro had undergone a change of heart about armed conflict against the government and this potentially so weakened the position of the drugs barons that they saw Pizaro as a threat to their authority.

During 1990 a Drugs Operational Command was established in the Pentagon to direct the US military involvement in the war against the drugs barons. A team of 30 military planners from the army's Southern Command worked with the Directorate of Narcotics and through this organisation with the Colombian authorities.

Since the US is the principal market for the Colombian narcotics traffickers, American involvement was inevitable. Patrols of well-trained American soldiers were leading anti-drug training companies of nationals in the jungles not only of Colombia but also in Peru, Bolivia and Panama.

By July 1990 the Barco government had seriously damaged but not destroyed Escobar's vast empire. The army and police had burned down dozens of cocaine laboratories, arrested hundreds of suspects, killed one senior drug trafficker and extradited 15 to the US. In response, Escobar began a terrorist campaign that led to the deaths of more than 300 civilians. The drugs war was obviously still raging when President Cesar Gaviria succeeded Barco as head of state.

The War in 1991

A hopeful sign came in May 1991 when the cocaine barons released one of their most famous hostages, Maruja Pachon, a television presenter, head of a national film company and wife of a senator. She was the ninth journalist to have been seized by the Medellin cartel. Her release was negotiated at a high price: the government agreed to abolish extradition to the US for drug traffickers, the punishment most feared by the criminals.[1]

By June 1991 remarkable developments had taken place in relation to narco-terrorism, guerrilla warfare, and corruption and inefficiency. Government contacts with the Medellin cocaine cartel appeared to be leading towards the surrender of the chief trafficker, Pablo Escobar, and up to 30 of his most notorious partners. A luxury jail was being prepared on a hillside outside Medellin.

A change was apparent in Colombia but, despite attempts at pacification, violent crime and deaths in armed conflict had increased by 16 per cent. A political pact was signed by the Liberals, the former rebels of M-19 and a radical conservative coalition. President Gaviria also signed the agreement which proposed that the Congress be suspended from 2 July, elections called for 6 October with the new parliament to meet by February 1992. In the meantime the President would rule by decree.

The hope was that followers of M-19's Antonio Navarro and of the National Salvation Movement of Alvaro Gomez would win a majority in the legislature. Minority groups protested that they had not been consulted and that the arrangements were a 'carve-up' to keep power for the old parties.[2]

Manuel Marulanda, known as Tirofijo or 'Deadshot', was asked to address the National Assembly. Since he leads FARC, whose guerrillas have been fighting in the hills and jungles since 1951, this was an astonishing invitation. It also recognised the reality that FARC cannot be politically ignored. Marulanda's second-in-command, Alfonso Cano, is chief negotiator for the combined forces of FARC and the ELN. Difficulties were caused by the rebels' demands to be recognised under the Geneva Convention as a legitimate belligerent force. There followed semantic disputes over the terms 'prisoners-of-war' and 'hostages' but at least by the middle of the year the parties were moving towards agreement and reconciliation. Just which side or which individual would turn out to be the 'winner' was doubtful. After all, the notorious Ochoa brothers, members of another Medellin cartel which surrendered in 1990, were still running cocaine shipments from their high-security cells.[3]

As so often in Colombia, hope turned to despair. Early in July, FARC and ELN fighters began a campaign of violent disruption. They blew up electricity pylons, oil pipelines and bridges, halting economic activity in many parts of the country. The campaign began as a tactical show of strength, designed to secure a better bargaining position at the peace talks, but it quickly grew into something much more serious because of the actions of an ultra-radical wing within the ELN opposed to any kind of negotiation with the government.

Spokesmen for the guerrillas asked for a 10-day postponement of the next round of talks, due to be held in Venezuela. They said that the leaders could not safely return to their units to consult with their followers because of an army offensive. The government peace negotiator, Jesus Bajarano, declared the talks suspended indefinitely, and the high hopes of June rapidly diminished.

The government announced that because of damage to the economy caused by the many bombings of the main oil and gas pipelines and the destruction of oil rigs, pumping stations and drilling camps, the nation's economic growth rate was down by a full percentage point.

Despite intensified helicopter surveillance and infantry patrols the military proved ineffective in protecting the economic infrastructure. They also suffered constant casualties from ambushes and attacks on police posts. The only success in this period was the killing of FARC's Sixth Front commander, Miguel Pascuas, in Cauca Department.

The rebels' offensive produced great hardship for the very people they claim to be fighting for — the ordinary citizens. There was no power and factories were closed across most of the north coast. Crude oil spilling from dynamited pipelines polluted hundreds of miles of rivers, leaving thousands of fishing families close to starvation. Food supplies could not reach some regions because of destroyed bridges and roads.

Scores of thousands of Colombians marched in protest against the guerrillas in Villavicencio, Nieva and Pitalito, either shouting angry slogans or shuffling along in grim and desperate silence, carrying banners. There was little prospect that the protests would move the guerrillas or influence them to change their tactics. They had seen that the indiscriminate violence of the cocaine barons had terrorised the government into making concessions. They believed that by replicating this violence they, too, could force the government to agree to their demands. The guerrillas may have misjudged both the people and the government. FARC seems to have lost its public sympathy.

There were indications that the government was close to military action: the arrival in Colombia of US-made Bronco OV-10 counter-insurgency aircraft certainly seemed to presage tough army tactics. Used extensively in Vietnam, the Bronco is a 'guerrilla-killer', designed to fly low over rebel territory, ferreting out targets with its infra-red sensors for its rocket, cannon and

Colombia Guerrilla and Narcotics War

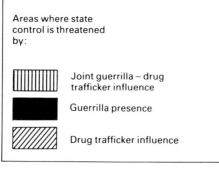

Areas where state control is threatened by:

- Joint guerrilla – drug trafficker influence
- Guerrilla presence
- Drug trafficker influence

machine-gun systems.

Self-Defence Groups

Paramilitary self-defence groups were first formed in 1987 by leaders of the community in certain prosperous areas, notably the cattle-breeding regions. Ironically, they had the backing of the army and Pablo Escobar, both of whom were eager to crush the guerrillas. The drugs cartel was investing some of its profits in ranches at the time and it was ready to help kill or drive out the guerrillas, who frequently kill or kidnap cattlemen. British and Israeli mercenaries were brought in to provide military training for the self-defence units.

A prominent leader of the paramilitary organisation was Henry Perez. Early in 1991 he was sentenced *in absentia* to 20 years imprisonment for his part in three massacres in which more than 50 people were butchered. The authorities made only a token effort to capture him because of his valuable help in the fight against the Communist guerrillas and against the drug chiefs.

On 20 July 1991, Perez was murdered in the lawless Magdalena Medio area. Other paramilitary leaders blamed Pablo Escobar, who was behind bars in his luxury prison. It was well known that Escobar had switched alliances. He was now fighting against his former cronies and helping to arm the rebels. The murder of Perez was very public. He was killed at Mass by men with machine-guns who sprayed the church with bullets, killing eight other people, including three children. The style of execution was typical of the drugs cartels.

Ariel Otero, who succeeded to the command of Perez's organisation, said that his forces would take armed action against Escobar, even in the security of his prison. While a few of Escobar's men have surrendered most of the cartel's military structure is intact. Only one weapon was surrendered — Escobar's pistol.

And Now — Heroin

By 1991 more South American cocaine was being manufactured than could be profitably smuggled. In short, the market for cocaine was saturated. A pound of pure cocaine was selling in Bogota for as little as the equivalent of $1,000. The authorities had known that drugs traffickers were switching to heroin production but only recently has the size of the opium poppy plantations been accurately assessed.

Colonel Rosso Serrano, commander of the Police Narcotics Brigade, took three companies of police into the mountains of the southern department of Huila on 'Operation Centenary Seven'. When it ended 2,300 acres of poppy plantation had been destroyed. This was much larger than the anticipated amount and is taken as an indication that the switch from cocaine to heroin is well advanced. Plantations have been reported in several other provinces.

The operation in Huila produced evidence that the heroin-makers are collaborating with FARC guerrillas. A rebel guardpost covered a fork in the trail leading to two large plantations. Documents inside the post showed that it belonged to the 29th Front of FARC.

Some poppy-growers have links with a Sicilian gang based in Venezuela. The profits are immense. High quality heroin sells at $40,000 a pound. A good worker in the opium growing business can collect an ounce of pure opium in a day. The worker is paid $6 a day, a little more than the industrial wage and much better money than can be made by growing corn or potatoes.

The next phase of Colombia's war will be waged against the heroin barons, who might well pay the guerrillas to protect them and their plantations. This will be an unholy and formidable combination.

Certain key questions arise from the Colombian drugs war and an attempt to answer them may clarify the government's inability to defeat the warlords.

- **Why can't the 200,000 members of the Colombian armed forces and police defeat the cartels?**

The law-enforcement organisations were never designed to be a narcotics strike force. Even at the end of 1990 the army, navy and air force were still mainly structured and equipped to repel foreign invaders, not terrorist nationals. The air force has concentrated on jet fighter squadrons, which have no role in modern South American politics, when it really needs helicopters for patrolling the jungles. The navy spent $90 million on repairs to submarines. It should have been buying or building light powerboats to run-down the traffickers on the country's rivers. Better co-ordination was also needed. At one time troops were closing in on the warlord Jose Gonzalo Rodriguez Gacha when an A-37 air force reconnaissance alarmed Gacha, his men discovered the army's approach and Gacha escaped. (He was killed some months later).

- **Will the capture and imprisonment of Escobar end the drug trade?**

This is impossible because Escobar, though the arch-criminal, is not the only dominant chief. A ring in Cali, which controls the flow of cocaine into New York city, operates almost with impunity because it never uses terrorist tactics. As a result, the security forces leave it alone. In any case they need this ring — it provides information about its Medellin rivals. Even if Escobar were to die, cocaine production would not slow down. Following Rodriguez Gacha's death, younger, even more vicious men moved in to fill his shoes.

- **Are the Colombian police and army corrupt?**

Most are not but many definitely are. Colombian officials admit in private that the army and, to a lesser degree, the police, are infiltrated by the drug gangs. President Barco on at least three occasions has ordered an investigation of suspected officers. Some were imprisoned for complicity with the drugs cartels. One senior officer, from his privileged position in Intelligence, had been warning the cartel of operations being planned against them.

- **Do the Colombian authorities really want to destroy the cartels?**

The answer is no. The objective is primarily to drive them out of Colombia and this would not necessarily curtail narcotics production. Officials distinguish between drug trafficking, which mainly affects and threatens the consumer countries, and narco-terrorism inside Colombia, which they are determined to stop. Many Colombians of all classes and in all walks of life call for negotiation with the cartel.

The War Goes On

While Communism was in retreat in most parts of the world the National Liberation Army (ELN) emerged from the forested hills on 29 January 1992 to make its boldest attack ever on an urban target. The guerrillas blew up three oil pipelines on the edge of Barrancabermeja, centre of the national petroleum refinery. Their aim was to bring oil production to a stop and they succeeded.

Oil gushed into the streams that supply drinking water to the sprawling *barrios*, the slum areas where the rebels constantly try to incite popular revolt. In fact, the ELN attack and its consequence turned the poor of the *barrios* against the ELN. Commander Pompilio, head of the ELN's urban operations, told a Colombia reporter that perhaps only ten per cent of the people understood the ELN's strategy. 'We want to strike back against oppression and protest about living conditions,' he said. 'We like to blow up pipelines but this is a means to an end.'

Despite the animosity the ELN aroused over the oil pollution, the fact is that it attracts recruits among unemployed youths. The ELN, reliably estimated to number 10,000 fighting members in April 1992, intends to carry the war into the cities and scorns the rebel groups that have laid down their arms. Late in March, ELN leaders rejected a cease-fire until Colombia adopts social and land reforms based on those in Cuba and Vietnam.

Throughout the first half of 1992, the rebels, every day, attacked army posts and the oil pipelines and killed or wounded suspected informants. In Barrancabermeja, Right-wing death squads hit back, thus increasing the violence. Most of the death squads' victims were suspected rebel sympathisers, such as union organisers.

The End of Escobar

In July 1992 Pablo Escobar escaped from his farcical imprisonment near his home town of Envigao and over the following year he eluded the Search Block unit and its 1,500 members that had pursued him. At least four times, minutes before a trap sprang shut, he slipped away and vanished. On 11 October 1993 eight members of the Search Block unit were sure that they had Escobar surrounded in a remote farmhouse in the Medellin district, but again he escaped.

On that occasion he left behind some correspondence, including a letter to his mother in which he said that he was tired and willing to turn himself in but believed that the government would not accept his surrender. He had made a standing offer of £15,000 for each Search Block officer killed but his own lieutenants were the ones who fell; Search Block killed 26 of Escobar's closest collaborators, including his brother-in-law. Mrs. Escobar and her children fled to Germany in mid-December but were at once deported back to Bogota.

Escobar called a Medellin radio station to complain about the German government's behaviour and then telephoned his family at their hotel room. The calls were traced through a scanning operation which pinpointed a house in west Bogota. In the raid that followed Escobar and his bodyguard were shot dead. The balance of deaths lay with Escobar: the violence of the past 10 years had cost Colombia the lives of an attorney-general, a justice minister, three presidential candidates, more than 200 judges, 30 kidnap victims, dozens of journalists and about 1,000 police officers.

The general public was jubilant over Escobar's death but the greatest celebrations were held in the city of Cali, where rival drugs barons gathered at a party arranged by cartel ruler Miguel Rodriguez Orejuela. The Cali cartel had already seized much of Colombia's cocaine market from Escobar's weakened Medellin organisation.

President Cesar Gaviria Trujillo announced to the nation: 'Colombia has shown that no criminal organisation can defeat the nation. We have won the war against Escobar and we will win any further wars.'

This was an implicit admission that further wars would occur. Diplomats in Colombia are unanimous in predicting that the Cali cartel will be no less vicious in its efforts to control a monopoly that brings in $9 billion a year. By mid-April, four months after Escobar's death, the Cali barons were negotiating with the surviving members of Escobar's Medellin network. According to an official of the US Drug Enforcement Administration, the Cali-Medellin union will create a super-cartel which could quickly become the most powerful criminal organisation in the world. Three new leaders to watch for in a continuation of Colombia's wars are: Miguel Rodriguez Orejuela, Jorge Luis Ochoa Vasquez and Jose Santacruz Londono. Orejuela is the leader of the Cali group; Londono is the brains of that group; Vasquez is a senior Medellin drugs chief.

The death of Escobar, at the age of 44, had no effect on Colombia's other conflicts but the prestige of the army and police became greater.

References

1. Maruja Pachon was watched constantly by four young men whom she described as 'very disciplined, trained killers'. They killed two other women hostages. Her release was a gesture supposed to demonstrate the drug barons' sincerity in the peace negotiations then under way. Cynics say it was a cold-blooded ploy designed to elicit sympathy for their cause from a nation desperate for signs of a truce.
2. Diplomats report that this view is realistic rather than cynical. The older parties have shown that they have staying power and therefore must be listened to. 'It is very much a case of the devil you know being less frightening than the one you don't know', a Scandinavian diplomat said. 'The problem for Colombia has been the appalling violence and the government will make any sacrifice to reduce it, even by making deals with its enemies.'
3. Nobody expects the drugs barons to serve more than two or three years in their luxury prisons. 'Pablo Escobar consented to being put in prison because he was promised he would be detained for only a short period and that while he was inside his life would be much the same as outside', an American official said.

Major aspects of the Colombian War dealt with in *War Annual 5* include:
    American military involvement.
    The war in statistics.
    A description of the Elite Force.

East Timor Resistance War

INDONESIA'S UNRELENTING OPPRESSION

Background Summary

In 1975 the Portuguese abandoned East Timor (also known as Timor-Dili) after 500 years of colonialist rule. A civil war followed, won by the *Frente Revolucionara de Timor-Leste Independente*, commonly known as Fretilin. Fretilin declared Timor to be an independent state on 3 November 1975 but Indonesia, which already possessed West Timor, ignored this and annexed East Timor. Within 10 years, the Indonesian army had massacred 200,000 of the 1975 population of 688,000.

Indonesia has the world's largest Muslim population — 176 million. From the outset, the Islamic fundamentalists among Indonesia's leaders told the army-of-occupation in East Timor that it was engaged in a *jihad* or holy war against infidels, as the Timorese people are Christians. Operating from bases in the mountains and forests, the Fretilin resistance fighters have raided Indonesian camps and posts. Unable to catch many of the guerrillas, the army takes reprisals against villages suspected of helping them.

The East Timorese people have many friends abroad, especially among Christian organisations, but they have been unable to bring real pressure to bear on the oppressive military government. A brief visit by the Pope in 1990 became a major event in the war of independence but after his departure the army crushed the people's demonstrations. Scores were arrested, beaten and tortured.

Despite the presence of 10,000 troops in East Timor, including units specially trained to deal with guerrillas, the Fretilin fighters often take the initiative. During 1990 they attacked army positions in the towns of Vieque, Meli-Meloi, Monumento and Usimur.

The resistance has been led by a Revolutionary Council of National Resistance (CRRN), of which Xanana Gusmao was the head. Gusmao was also leader of Fretilin and commander of Falintil, the armed wing of Fretilin. The CRRN was composed of the central committee of Fretilin and the Falintil high command.

In the 1980s the resistance recruited more non-Fretilin members, necessitating changes in the command structure. Gusmao left Fretilin, though not CRRN. The latter was reorganised into the National Council of Maubere Resistance (CNRM), which has a broader membership than before. Falintil is no longer Fretilin's army but a national army. Xanana Gusmao is head of CNRM and commander of the non-party Falintil.

The War in 1991

In August 1991 Amnesty International, the human rights organisation, produced another in a series of reports on East Timor. It accused Indonesia of maintaining its ban on observers visiting the country to examine charges that political detainees had been tortured and executed. The government was refusing to allow Amnesty officials into the country, despite Indonesia's stated decision to play an active role in the international human rights community.

'The continued denial of access raises serious questions about the government's continued commitment to protecting and promoting human rights', stated the report. This commitment could not be taken seriously until it investigated allegations of violations of human rights, published its findings and brought offenders to justice.

Amnesty was particularly concerned about what it called an accelerating pattern of short-term detention, torture and ill-treatment of political opponents. It estimated that at least 400 people had been detained since 1988. Most had not been formally charged or tried during detentions that ranged from a few days to several weeks. The report mentioned Amnesty fears that hundreds of people who had 'disappeared' in the territory since 1975 may have been killed, including at least 30 executed by the Indonesian security forces in 1990 and early in 1991.

A Portuguese parliamentary visit to East Timor was arranged for November 1991. In August the Indonesian military began to round up the imprisoned dissidents in anticipation of this visit, the objective being to discourage Fretilin from 'exploiting' the first Portuguese visit since 1975.

An army document sent by the commander of the Special Operations Executive in East Timor (known as *Pangkolakops*) to the director of the Strategic Intelligence HQ, was leaked to the East Timor News Agency. In military code, it describes operations to stop and destroy the Fretilin guerrillas and analyse the preparations for the visit.[1] It also stated that the special branch of strategic intelligence was to investigate the escape in May 1991 of four Timorese soccer players who were taking part in the Arafura Games in Darwin, Australia.[2] The special forces were ordered to 'uncover and deter' the underground networks involved with the preparations for the visit or with the Fretilin guerrilla movement.

The Indonesian ambassador to Australia, Mr. Sabam Siagian, said that the reports were a fabrication but a spokesman for the Roman Catholic archbishop of East Timor, Bishop Belo,[3] confirmed the reports and said that the East Timorese were 'living a time of terror'. Just before the Bishop spoke, two East Timorese youths had been shot dead by troops near the Motael Roman Catholic Church in Dili. As is their custom, the occupation authorities blamed the shootings on 'anti-integration forces'.

In the event, the Portuguese government cancelled the visit of its parliamentary committee after the Indonesian government vetoed the entry of the Lisbon-based Australian journalist Jill Jolliffe. The Indonesians claimed that she was a 'crusader' for Fretilin. The journalist is certainly a specialist on East Timor, with sympathy for the plight of the East Timorese, but she is also very much the professional observer and commentator. The real concern of the Indonesians is that this particular observer, able to communicate with the people in their own language and with a thorough knowledge of the 16-year war, would find out the truth of the military occupation.

Following this breach in relations, the UN Secretary-General, Javier Perez de Cuellar, appealed to the two countries to reach an agreement on the visit. He said: 'Both States should reconsider their positions in view of the efforts invested by all sides in arranging such a visit and the extensive preparations that have already taken place.'

The Dili Massacre

On 28 October 1991 a young Timorese independence activist, Sebastiao Gomes Rangel, was shot dead by Indonesian troops in the compound of the Dili church. Several hundred Timorese gathered at the church at 6.15 am on 12 November for a memorial mass, and then marched through the town to the cemetery where Rangel had been buried.

The unarmed and peaceable crowd had been in the cemetery for ten minutes when, at 8 am, a large contingent of soldiers approached from two directions. Without ordering the crowd to disperse, one group formed a line twelve abreast and opened fire. Among the foreigners present was a cameraman from Yorkshire Television who recorded much of the carnage; he smuggled his film out of the country and it was later widely distributed.

Amnesty International claimed that between 50 and 100 people died in the massacre. Sources among the Timorese emigré population in Australia claimed the death toll was between 125 and 150. The Indonesian army put the figure at 19 but admitted that 89 civilians had been wounded. Subsequent inquiries by diplomats and churchmen in Dili established that more than 90 people perished in the shooting while others later died of their wounds.

The army's regional commander, General Rudolf Warouw, claimed that during the march from the church to the cemetery a senior officer had appealed for calm, only to be attacked and seriously wounded with a machete. Knives, rocks, pistols and rifles allegedly seized from the 'rioters' were displayed for the foreign press. However, no weapons were seen by the foreigners who took part in the march nor were they in evidence in the Yorkshire Television film. The only 'incident' was a brief scuffle between a few demonstrators and soldiers en route to the cemetery but nobody had been hurt.

International reaction was swift and critical. The European Community 'vehemently condemned' the killings; the US Administration viewed the event 'very seriously' and the Australian Prime Minister Bob Hawke, called the Indonesian military response 'tragically excessive.'

The chief of Indonesia's armed forces, General Try Sutrisno, defended the troops' actions in a speech to graduates of the National Defence Institute the following day. 'Come what may, let nobody think that they can ignore ABRI' (Indonesia's armed forces); the general said. 'In the end they will have to be shot down.' His speech was reported in the Jakarta daily *Jayakarta* on 14 November.

The Dili massacre threw Indonesian politics into turmoil. The National Investigative Commission was forced to examine the evidence and it produced a relatively credible report. It was certainly more frank than critics of Indonesia's policy in East Timor had anticipated, if only because there were compelling financial reasons for Jakarta to handle the Dili killings inquiry in a

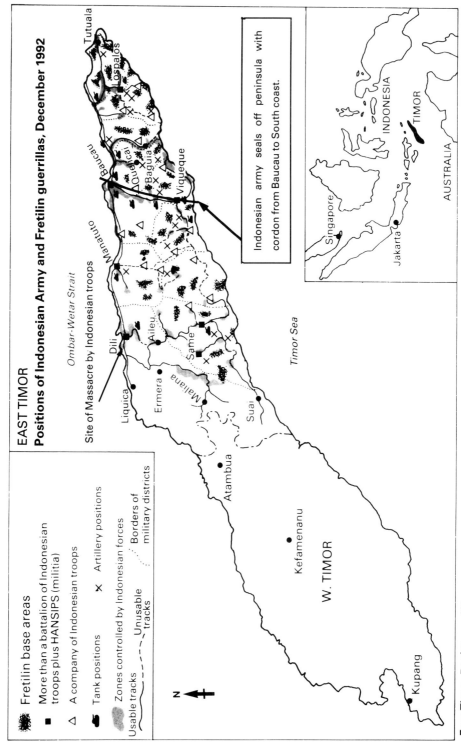

East Timor Resistance War

way that was internationally acceptable. General Warouw was dismissed from his command in East Timor.

While the world publicity may have helped Fretilin's cause the movement suffered a setback in January 1992 when its deputy chief, Jose da Costa (also known as Mauhudu) was caught by troops while he was lying low recovering from malaria. There followed an even greater disaster when Xanana Gusmao, the resistance chief, was captured.

The new army commander in East Timor, General Theo Syafei, announced that during interrogation da Costa had named 40 'accomplices'. More than 50 East Timorese were arrested in the weeks following Syafei's appointment. Later, some were gaoled on charges of passing secret documents to the exiled independence leader Ramos Horta, who lives in Australia.

Damage Limitation

During 1993 and 1994 the Indonesian government was engaged on a protracted damage limitation exercise following international unease about the Dili massacre. For instance, Xanana Gusmao was shown on film apparently relaxed in a comfortable prison. Numerous Indonesian public figures discounted or denied army maltreatment of the Timorese people and some foreign journalists were allowed to visit East Timor, though only under strict control.

Propaganda aimed at discrediting those people who condemn the invasion of East Timor and the repressive military occupation has been largely successful, though some prominent Indonesians have publicly contradicted the official line. Australia permitted Indonesian army personnel to train in Australia with units of the Australian army and government spokesmen have been conciliatory in tone towards Indonesia. This is understandable in view of the vast and valuable market that Indonesia represents in Australia.

However, the facts remain: East Timor was illegally and forcibly invaded and occupied against the wishes of its people and the resistance movement continues to struggle. Many international observers say that East Timor is a more deserving case than Bosnia and Somalia.

The Real War

The new resistance commander, Nino Konis Santana, aged 39, claims effective control of East Timor. According to this able leader, a former schoolteacher, the Indonesian army can control an area only during a military operation and when it finishes the people continue with their lives.

The task of the resistance is not to defeat the enemy militarily but to deny them the ability to integrate East Timor into Indonesia, according to Santana. To achieve this limited objective, the guerrillas are attacking and ambushing the Indonesians more frequently, while at the same time reorganising their self-defence. The damage done by the capture of Xanana Gusmao has been repaired.

Fretilin's performance has been all the more impressive because General Syafei has been on the offensive in the entire country. His objective is well known: extermination of the guerrillas, both physically and through desertions.

Yet resistance is stronger than ever. A Timorese priest, Father Suarez, said 'After 12 November 1991 the form of resistance changed dramatically. It increased the deep-seated hatred they feel towards the Indonesian presence in East Timor, even while the Indonesians continue with their acts of terror'.[4]

A surge of recruitment in the guerrilla ranks is evidence of Fretilin's increasing strength. Many of the new recruits had surrendered in earlier years to the Indonesians. These men run a great risk because they know that the Indonesians will butcher them should they be caught.

Fretilin's organisation in 1994 is interesting. It has about 800 full-time fighters in units which cover all of East Timor. Backing these 'regulars' are reserve groups, numbering from 230 to 1,500, in specified regions. These men are available at very short notice, which they generally received by coded radio messages. Within the settled areas are tens of thousands of organised activists. Santana says that Fretilin could readily raise 100,000 armed men if the weapons were available. Lack of arms is the guerrillas' weakness, thought there is evidence that some supplies become available through corrupt or sympathetic Indonesian army officers.

References

1. The document was leaked by a military source. This might in some circumstances arouse suspicions that the army was involved in disinformation. However, the embarrassment and anger of the Indonesian High Command was very evident. A subsequent investigation failed to uncover the official responsible for the leak.
2. The four Timorese asked for political asylum. Relations between Australia and Indonesia have been strained by the murder of Australian journalists in regions controlled by the Indonesian army. Over the years since 1975 some Australian governments have been remarkably co-operative with Indonesia; others have been more forthright in their condemnation of human rights abuses.
3. Bishop Belo, head of the Roman Catholic Church in East Timor, has emerged as an heroic figure. He has spoken out in defiance of Indonesian authority. 'Everyone is obliged to think and do according to the Indonesian political system', the Bishop said. 'They are constantly told that they are an integral part of Indonesia and that there is no other future for them.' The Indonesian government has pressed the Vatican to remove Belo but he is protected by his increasingly international prestige.
4. The priest was talking to Max Stahl, the British film-maker who filmed the Dili massacre in November 1991. Stahl published an article in the *Sydney Morning Herald* of 15 February 1994 entitled 'Guerrillas in the mist: the war of resistance in East Timor'.

War Annual No. 5
    carried a particularly detailed account of the East Timor war. Included are details of changes in Indonesian Army structure; the first published report of the *Kapan Pulang* campaign; the operations conducted by Colonel Prabowo; foreign political support for the Timorese; the visit of the Pope.

Ethiopia — Eritrea — Tigre — Somalia

Background Summary

The root cause of the conflict in many regions was Ethiopia's refusal to grant independence to its minorities. The predominantly Muslim Eritreans claimed that they should have a separate state from the mainly Christian Ethiopians. Ethiopia annexed Eritrea in 1961 and an army of 80,000 occupied the area. The Eritrean Liberation Front (ELF) fought off the Ethiopians but the Soviet Union then backed Ethiopia as the best prospect for dominating the Horn of Africa. With massive Soviet support, the Ethiopians drove the ELF from parts of Eritrea.

The Tigrean People's Liberation Army (TPLF) was also fighting for autonomy, while in the south the Ogaden tribes struggled for reunion with their countrymen in Somalia. The ELF (which changed its name to Eritrean People's Liberation Army — EPLF) and the TPLF were regular armies by 1986 and the EPLF fought from a trench line of 340 miles from the Red Sea to the Sudanese border. Both armies won remarkable victories in 1988, 1989 and 1990. Details are given in *World in Conflict — War Annual Nos 3, 4 and 5*.

The great battle of Massawa in 1990 was a devastating defeat for the Ethiopian forces and the long war tilted decisively in the Eritreans' favour. On the southern front in April 1991 they overran several garrison towns on the Asmara — Addis Ababa road. In May 1990 President Mengistu and his ruling group, the Dergue, called for 'total mobilisation' in a desperate attempt to shore up his defences and increase the size of his army from 350,000 to an impossible 5,000,000.

The core of Mengistu's defence against the hatred of his people was a specially trained bodyguard of 2,500, many of them tall Southerners with ritual scars on their foreheads. These guards were said to be illiterate and Mengistu kept them happy and discontented in turn, giving them more privileges than regular soldiers while encouraging rivalries among ethnic groups. The presidential guard, known in the army as 'the fattened regiment', guarded Mengistu's palace in Addis Ababa, which was ringed with tanks and anti-aircraft guns.

By the end of 1990 it was obvious to diplomats and observers in Ethiopia that Mengistu and his army could not hold out for much longer. The rebel groups were winning one battle after another, the army's morale was in shreds and Mengistu made matters even worse for himself by executing 'failed' generals. This left him without his most experienced field commanders and discouraged others from taking risks. Many deliberately manoeuvred their troops to avoid battle.

When the Ethiopians *were* brought to battle, they often fought bravely but by 1990 Mengistu's motivating propaganda and rhetoric was wearing thin.

Another distressing factor for Ethiopian soldiers was knowing that many thousands of their comrades were languishing in captivity, completely without contact with their families. According to Eritrean leaders, many Ethiopian soldiers feared being taken prisoner — even though they knew they would not be tortured or massacred — more than being killed in battle. They knew that captivity meant apparently endless years behind barbed wire on a barren hillside, where they would be perpetually hungry. Exposed to the elements, they would be cold in winter, and without shelter from the fierce sun of summer. They also knew that Mengistu had given orders that soldiers who had been captured must never again be mentioned in the army, especially not in casualty lists. They normally had little opportunity to write home and when no further letters arrived, their families could only assume that their men were dead. In fact, Mengistu several times said that it was more honourable for an Ethiopian soldier to be dead than taken prisoner. Throughout the war there was no organisation to speak for the captured soldiers. Neither side permitted the Red Cross to send its field officers to prisoner-of-war camps.

The War in 1991

In some of the last actions of the war the EPLF forces progressed further south than ever before to liberate the harbour towns of Berasole and Beylul at the beginning of April. In pitched battles between 4 and 6 April the EPLF units routed two élite 'Spartakus' brigades, which had been trained by North Korean instructors, a 'special' commando brigade and four artillery battalions. Total enemy losses were 2,700 dead or wounded. The EPLF destroyed three tanks and captured quantities of light and medium arms. Their own losses are believed to have been fewer than 100. Berasole is 70 miles and Beylul only 30 miles north of Assab, Ethiopia's last stronghold on the 600-mile Eritrean coastline. Ethiopian attempts to recapture Beylul were repulsed and, according to EPLF estimates, the Ethiopian army suffered another 2,200 casualties in mid-April.

In the meantime, forces of the Ethiopian People's Revolutionary Democratic Front (EPRDF) had victories in Welega and Shewa provinces. EPRDF units captured Bimbi, Mejo, Fincha and Ambo in April. Mengistu's troops retook Ambo but held it only briefly. As April ended almost all of northern Ethiopia was in rebel hands and Mengistu was clearly finished.

The three main groups now fighting the government on this front were the EPRDF, an allied group of Eritreans fighting for independence and a smaller band of Oromos. They were not eager to storm Addis Ababa, knowing that they would suffer heavy casualties. As a result, the US government tried to arrange a peaceful transfer of power to a broad-based transitional government that would rule the country until elections could be organised. Mengistu claimed that he alone represented unity for Ethiopia against the secessionist demands of the Eritreans. The American negotiators argued that if there were no political settlement the Eritreans would win their independence by force.

Before fleeing to Zimbabwe, Mengistu nominated Tesfaye Gebre-Kidan as his successor but his regime — if it could be called that — did not last long. Government troops, demoralised and poorly led, turned on one another as

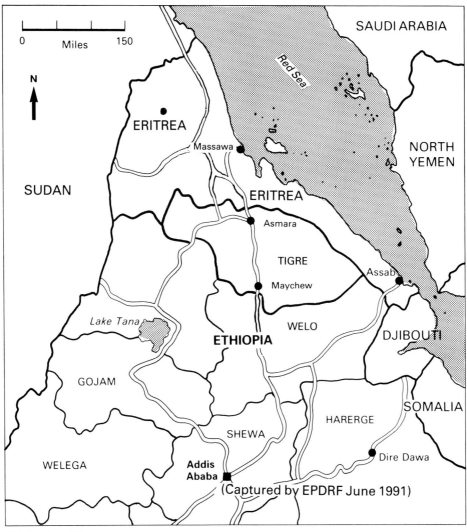

Ethiopian Liberation Wars at an End

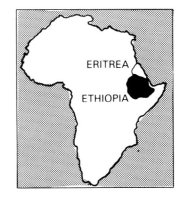

Ethiopian Tribal Areas

units formed factions of their own. State property was wantonly looted and many people were killed in random shooting. Tesfaye summoned the US Chargé d'Affaires in Addis Ababa to tell him he could no longer control the situation. The interim Ethiopian leader promised he would issue a unilateral ceasefire and tell the people of the capital to welcome the rebels into the city. He did not fulfil the second part of his promise but he did proclaim a ceasefire before seeking asylum at the Italian embassy.

Herman Cohen, US Assistant Secretary of State for African Affairs, announced in London that America was 'recommending' that the EPRDF should enter Addis Ababa quickly to 'stabilise the situation'. The Front did so but United States' encouragement of the group made it the target of much animosity in Addis Ababa.[1] When Cohen announced that the US supported the Eritreans' right to self-determination, mobs of Ethiopians marched to the gates of the American Embassy, shouted slogans and threw stones into the compound. Demonstrators dubbed the change of government as 'Cohen's coup'.

THE MAIN GROUPS AFTER MENGISTU'S FALL

Eritrean People's Liberation Front (EPLF)

The oldest and best organised of the insurgent groups, the EPLF was established in 1970 as a breakaway faction of the Eritrean Liberation Front, formed in 1958. It is a secular, multi-ethnic organisation devoted to independence for Eritrea province. Originally Marxist–Leninist, since 1985 the group has embraced the concepts of a regulated market economy and political pluralism. With 95,000 fighters, the group is led by Secretary-General Issias Afewerki, a Christian aged 45. It has received support from Kuwait, the United Arab Emirates and Syria.

Ethiopian People's Revolutionary Democratic Front (EPRDF)

A four-group umbrella organisation formed in 1988, the EPRDF often known simply as the Democratic Front, is dominated by the Tigré People's Liberation Front. The group, founded in 1975, originally campaigned for autonomy for Tigré province. Following Mengistu's downfall, the Tigreans say that they favour a unified Ethiopia, though they support the demand of their military ally, the EPLF, for a referendum on independence for Eritrea. Led by chairman Meles Zenawi, aged 36, the Tigreans, once firm Marxists, now preach an odd mixture of Communism and democracy. The Democratic Front comprises 80,000 orderly and disciplined guerrillas and has been supported by Libya and Sudan.

Oromo Liberation Front

Established in 1975, the OLF demands either autonomy or independence for the southern provinces. This region is the heartland of the Oromos, also known as the Gallas, who complain that they have been oppressed by the politically dominant Amharas. The Oromos are Ethiopia's largest ethnic group, with 40 per cent of the population of 51 million, but their power has been diluted by their dispersal throughout the country. As a military force the OLF is marginal with only 7,000 fighters. It is allied with and supported by the Eritrean rebels but it fears the Tigreans. Led by Secretary-General Galasa Dabo, the Front has been aided by Sudan.

Opposition to the EPRDF resulted from the eccentric politics of the group, which is an umbrella organisation of resistance factions dominated by the Tigrean People's Liberation Front. Originally rigid Marxists, the Tigrean fighters proclaimed themselves converts to pluralism and the free market. The policy statements of the EPRDF, formed in 1988, contain elements of old orthodoxy. The people of Addis Ababa do not understand their professed moderation, they think of the Tigreans as a group more Marxist than Mengistu himself.

Ethnic tension was a central element of the trouble in Addis Ababa in mid-1991. The central government, like the capital itself, has long been dominated by the Amhara people, who consider themselves the most sophisticated of the Ethiopians and therefore the country's rightful masters. The Tigreans, who speak a different language and stem from a distant region, have been rivals of the Amharas for 2,000 years.

Problems of Peace

Between the departure of President Mengistu and the arrival of the new interim government much of Ethiopia was in chaos. In a week of lawlessness, hospitals, schools and stores were looted. Aid workers' depots were raided and destroyed by gangs of bandits and by crowds of ordinary people making the most of the opportunity to grab whatever they could before law and order were re-established. What happened to the Irish aid agency, Concern, was typical. From 11 stores for its famine-relief programme in north Omo, the agency lost 11,000 tons of food, enough to feed 16,000 families for a month. At Shone, where four aid workers were trapped in a newly-established compound, a crowd broke in and carried off the food, tools, refrigerator, computer, cooker, beds, doors, roofs, window frames and even the wiring.

The organised and disciplined troops of the EPLF and the EPRDF finally arrived in most areas and restored order. They seized thousands of weapons either abandoned or sold by fleeing soldiers.

The most urgent problem facing Ethiopia after the goverment's downfall was what to do with the large army, now leaderless and useless. Scores of thousands of former soldiers were expelled from Eritrea, where many had served for years, into Tigré. On foot, many without shoes, they poured over the mountains in rain and hailstorms.

In the northern town of Adigrat, 15,000 civilians from Eritrea were separated from the 13,000 ex-soldiers. All were of Ethiopian origin even though they lived in Eritrea but now none were welcome there. No solution was found. The large bodies of troops were penned into designated areas and lived there in great hardship until they were released in small groups to make their way back to the tribal lands. In many cases this meant a walk of hundreds of miles over a period of weeks. Some former soldiers were still in captivity at the end of 1991. The government feared that large numbers arriving in their homelands would be destabilising and that food riots would ensue.[2]

Under a charter adopted by the 81 delegates representing 24 different groups, the Eritreans, as well as dozens of other nationalities, have the right to self-determination and even succession. The delegates, who were, in effect, a quasi-government, agreed that in mid-1993 Eritreans would vote on whether to break away from Ethiopia.

Many non-Eritreans oppose the province's independence for economic as well as nationalist reasons. Without Eritrea, with its long Red Sea coast, Ethiopia would be landlocked. International food aid, which is often vital, enters the country mainly through the Eritrean ports of Massawa and Assab. The Eritreans have pledged that they will permit goods to flow freely through their territory but many Ethiopians doubt if they can trust such promises from a group that has fought the Addis Ababa administration for three decades.

The EPLF, under Issias Afewerki, faces new problems. It cannot afford to antagonise the newly-installed government in Addis Ababa. Nor can the Front alienate the international community on which it depends for economic aid and famine relief. The EPLF has to convert itself from a rebel army to a civilian government capable of reconstructing a region devastated by 30 years of war. Otherwise, the leadership runs the risk of splitting the unity that has brought the independence movement so far.[3]

Ethiopia's Naval Disaster

A naval operation provided an interesting postscript to the end of the Ethiopian war. As the conflict was ending, 12 ships of the Ethiopian navy fled from Assab to the Yemeni port of Mocha, about 220 miles south of San'a, the Yemeni capital. With their complement of 1,700 sailors, they were allowed to stay in Mocha without actually being interned. Within a week the fleet was followed by an Eritrean gunboat. Its crew blew up seven of the Ethiopian ships and damaged the others, but only one man, the Ethiopian naval commander, was killed. Others were wounded in exchanges of small arms fire. No Yemeni forces were in the area at the time, which indicates either that the EPLF's intelligence was good or that the Yemenis anticipated an attack and kept out of the way.

Following the EPLF raid the Ethiopian navy virtually ceased to exist. Its two Soviet-made frigates had been previously disabled at Massawa, the main naval base. As far as is known, the only remaining vessels of any importance are two small Soviet-made amphibious ships and some missile craft, which have been taken over by the Tigreans at Assab.

* * *

Ethiopia still has much strategic value because of its location on the Red Sea and its proximity to the Arab world but, like others in the Horn of Africa, it is no longer the geopolitical battleground that it was during the Cold War, when Washington and Moscow backed rival clients.

Mengistu as a Military Commander

While still in the army, Mengistu Haile Mariam reached the rank of Lieutenant Colonel in the army's supply department; he was, in fact, a senior quartermaster. Following his seizure of political power, he seemed to assume that he had a talent for military leadership and assumed command of the armed forces, though he applied no rank to himself. For some years he permitted his Generals to direct the nation's wars but lack of complete success exasperated him and in 1987 he began to plan campaigns and battles by himself.

It might have been expected that his training and experiences as a quartermaster would have given him an understanding of logistics, but he frequently asked for movements that could not be carried out in the time he demanded. At least twice he ordered corps to change position at the height of conflict, with consequent disruption and lack of control. On both occasions a disaster was averted only because of Generals' initiative and the rebel forces' lack of vehicle mobility.

Mengistu preached decisiveness to his senior officers and could indeed display this military virtue himself but he lacked the confidence to remain decisive. Having given an order he would rescind it. One of his Generals said that Mengistu lacked nerve, by which he meant the courage to persist with a chosen course of action.

Always heavily dependent on his massive stocks of Soviet armour and artillery, backed by a powerful air force, Mengistu could not understand the irregular tactics adopted by his more innovative opponents. Despite advice from his Generals, he never succeeded in finding a way of defeating the guerrillas who had turned themselves into armies. Even the better Ethiopian Generals and their Soviet advisers wasted prodigious quantities of ammunition on targets that did not exist. The EPLF and the TPLF did not concentrate their troops; by dispersal they denied the Ethiopian army an opportunity to attack a large and static force.

SOMALIA — END OF THE BARRE REGIME

The modern war history of Somalia began in 1969 when Major General Siyad Barre seized power in a coup. Almost at once he turned away from the

West and accepted aid and advice from the Soviet bloc. Still in need of support, Somalia joined the Arab League in 1974.

Barre launched an offensive against Ethiopia for control of the Ogaden region. This was a political-military blunder of startling proportions. The Soviet Union at once switched allegiance from Somalia to Ethiopia; that is, they backed President Mengistu instead of Siyad Barre as the prospective winner of the dispute. Barre expelled all Soviet-bloc military personnel and civil advisers and withdrew military facilities.

The smaller and weaker Somali army was pounded into defeat and returned home in disgrace. To survive, Barre surrounded himself with members of his own clan, the Darode, and his sub-clan, the Marehan, ruthlessly crushed all opposition and wooed the Americans. Despite massive aid from the West and the Gulf states, Somalia's economy was in the hands of corrupt incompetents and it deteriorated rapidly.

Somalia and Ethiopia remained in confrontation until May 1988 when Mengistu and Barre negotiated a reciprocal withdrawal, in effect a military stand-off. This provoked the Somali National Movement (SNM) into a civil war against Barre. The war was fought on tribal divisions. The large Ogadeni tribe made up much of the army, with part of the Darode clan. The Isaq tribe in the north and the Haweih in the south opposed the government.

By mid-1990 Somalia was in chaos. Many thousands of country dwellers found refuge from the fighting in the capital, Mogadishu, even though this city, too, was lawless. Barre was protected by his fiercely loyal Red Beret battalion but even it could not hold out indefinitely.

The War in 1991

The offensive which deposed Barre began on 30 December 1990. The recently-formed United Somali Congress (USC), drawing on the support of the rebel movements, the SNM and the Somali Patriotic Movement (SPM), decided on an all-out assault.

Barre found the Isaq-dominated SNM moving on Mogadishu from the north and the Ogadeni-led SPM sweeping in from the south. At first he tried deception and diversion. He promised a referendum on a new constitution that would lead to a multi-party democracy. When this ploy failed, he tried to win the rebel movements, one by one, into an alliance with him against the others. This divide-and-rule approach had worked for Barre before but now it failed.

As many of his soldiers deserted, particularly to the USC, Barre collected his loyal Red Berets, almost all of them from his Marehan sub-clan, and established one formidable redoubt at his palace, the Villa Somalia, and another at the airport. Well-armed and with plenty of ammunition, Barre's men opened the battle of Mogadishu with heavy artillery and machine-gun fire. In previous fighting, this simple tactic had been enough to win the day for Barre. On this occasion, the rebels were better organised and the need to repel a two-pronged assault halved Barre's available fire.

The battle was one of the fiercest seen in Africa for many years other than in Liberia in 1990, and even there no single battle lasted a month, as this one did. Ferocious hand-to-hand fighting took place in and around the two redoubts.

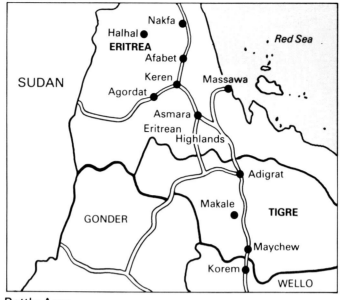

Battle Area

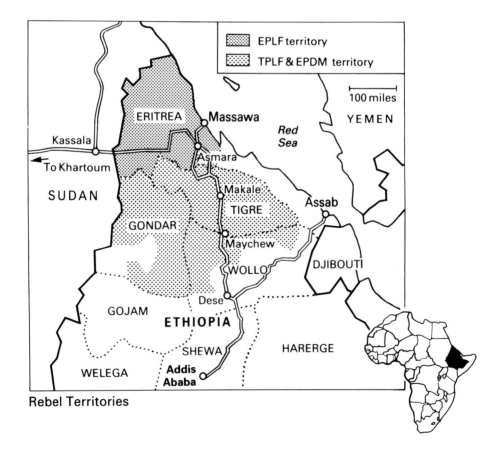

Rebel Territories

On 29 January the Villa Somalia was stormed. Rebel soldiers who had changed sides and a strange mixture of tribesmen and political opportunists, as well as criminals, burst into the palace itself. They hacked the defenders to death, looted the palace of everything that could possibly be carried and destroyed the rest. The airport redoubt also fell. Killing and looting went on in many parts of the city and an observer reported that the sky was black with vultures waiting to feast on the piles of dead bodies.

Barre escaped from Villa Somalia, a fact announced by the USC on 30 January. In fact, he had fled to Kenya.[4] The USC appealed to workers to return to their jobs and invited the rebel movements to join in forming an interim government under Ali Mahdi Muhammad as the new president and Umar Arteh Ghalib as a stop-gap prime minister. The USC also requested the international community to resume diplomatic relations with Somalia.

The new government faced the enormous task of trying to reconstitute a country which Siyad Barre had spent 21 years wrecking. Putting Somalia together again has proved impossible because of the horrors uncovered following Barre's downfall. Barre's troops carried out many massacres against the Isaq clan of the north in reprisal for an attack launched by the SNM in May 1988. Villages were denuded, as survivors fled to Ethiopia and only after the war ended could they return home. It was only then that the extent of Barre's oppression became known. The man actively responsible for much of it was the dreaded General Morgan, Barre's brother-in-law.

In Hargeisa, the capital of the north, the central committee of the SNM came under enormous public pressure to declare a separate state. The Republic of Somaliland, based on the former British protectorate, was declared, under the slogan 'No more Mogadishu'.

Hargeisa was a Beirut-like wreck. Between May and August 1988 government troops and the SNM fought a destructive battle here. Six MiGs and Hawker–Hunter aircraft sent by Barre repeatedly bombed and strafed the rebel areas and slaughtered refugees fleeing towards Ethiopia. According to some foreign aid sources, half the population of 500,000 were killed. With Barre's approval, his troops then stripped the city. When they had finished, hardly a roof, window-frame, door or piece of furniture remained. Even wiring and piping was ripped out, all to be trucked south or sold at the Ethiopian border.

To ensure that Hargeisa would not readily be resettled, the troops destroyed the water pumping stations and polluted the wells. Finally, they planted thousands of mines in the ruins and along the paths. Casualties from these mines during the first half of 1991, as refugees returned to the ruined city, were frequent. *Médecins Sans Frontières*, which had a Dutch medical team in Hargeisa, brought in bomb disposal experts from Rimfire International, a London-based security firm, to help co-ordinate mine removal.[5]

Barre as a Military Commander

Barre was commander-in-chief of the Somali forces when he mounted the coup which brought him to political power. From a humble background, he nevertheless became an inspector of police and later went to Italy to attend a

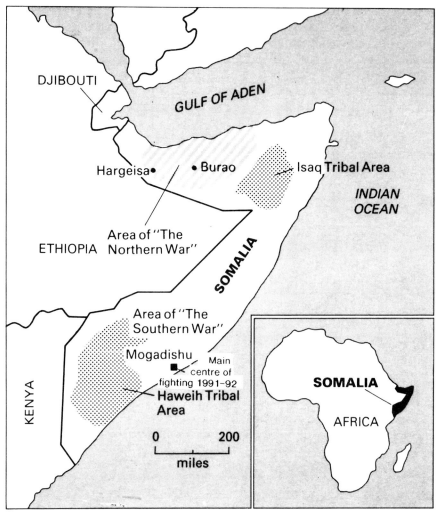

Somalia

military academy. He seems to have learnt little about strategy and tactics and even less about man-management. The Italian army could hardly claim him as one of its brightest products.

What he lacked in talent, Barre made up for in ruthlessness. He was brutal to all ranks and insistent on vengeance. He had seen his father killed by Isaq tribesmen in a clan feud and the memory never left him. The Isaqs had to be made to pay. Yet, by strange irony, he wanted to diffuse the clan orientation of Somalia and he brought in laws to prohibit clan or tribal politics.

An advocate of quick, decisive and lasting victories, Barre encouraged his senior officers to wipe out defeated enemies rather than capture and imprison them. In one sense, this policy paid off for when enemy soldiers saw defeat staring them in the face they often defected and offered to join Barre's forces. Those who were rejected were butchered anyway.

Barre's lack of strategic sense was shown by his attack on Ethiopia, in an attempt to capture the Ogaden region. There was not the slightest possibility that his army could be successful, especially in open country where it would be vulnerable to attack by Ethiopia's large air force.

A megalomaniac and introvert, Barre kept all military planning and personnel appointments in his own hands. He had the benefit of some able senior officers but he rarely, if ever, called for their advice and they knew better than to volunteer it. Barre was so hyper-suspicious that he always imagined an ulterior motive when any of his officials proffered some suggestion. Barre executed his most able generals and brigadiers because he thought they were too soft, and appointed his brother-in-law, the atrocious General Morgan, to the key post of governor of northern Somalia. He was, in effect, giving Morgan a licence to annihilate the Isaq clan and Morgan did his best to fulfil Barre's wishes.[6] Ultimately, Barre and his gang destroyed Somalia.

The War in 1992

The end of the Barre regime was the signal for the beginning of a new phase of the civil war. The large Hawiye clan split into two main factions, one to follow the Interim President of the country, Ali Mahdi, the other joining the ranks of General Muhammad Farrah Aidid's faction. From December 1991, fighting was virtually incessant and savage, with civilians being killed or wounded in their thousands.

In January, UN Security Council Resolution 733 called on both sides to respect a ceasefire so that increased humanitarian aid could be sent into Somalia. Neither provision of the resolution was adhered to. Somalia has an abundance of arms and ammunition stockpiled during the Barre regime's government but, in any case, it is not possible to close Somalia's long borders against embargo-breakers. Large supplies were arriving from Libya in 1992 as Colonel Gaddafi sought to destabilise the country still further.

On 14 February, ceasefire talks began at the UN in New York. The UN Secretary-General Dr. Boutros Ghali had invited both factions to send three representatives to New York to try to negotiate a ceasefire with the help of officials from the Arab League, the Organisation of African Unity and the

Islamic Conference. Fighting continued throughout the preliminaries to the meeting, which was abortive.

At the end of March, the UN estimated that 30,000 people had died in the civil war while another 1.5 million faced famine whose severity was described by Africa Watch as 'unequalled in Somalia's history.' About 250,000 people had been driven from their homes and an estimated 70 per cent were suffering from malnutrition.

Neither side had any sense of strategy. Their objective was merely to kill and destroy. Very young men whose allegiences were not apparent carried a great variety of weapons. When the army barracks were looted, everybody obtained a rifle, machine-gun or rocket-launcher. The presence of roughly equal numbers of AK-47s and M-16s illustrates the country's fluctuating political relationships during Barre's 20-year rule. The young fighters' preferred vehicle was the pick-up truck, with a 50mm machine-gun mounted on the back. Many of them moved in gangs, loyal to nobody but themselves, preying on the weak, looting at will and killing at random.

United Nations Intervention

The mounting conflict in Somalia, the deaths of innocent people, international clamour for the UN 'to do something' and the fear that if nothing were done the unrest could spread into other parts of Africa, drew the UN reluctantly into Somalia. The US, with equal reluctance, was the country on to which the main military responsibility fell. Pakistan, Nigeria and Australia contributed smaller forces to the UN's peace-keeping operation. The precise objects were to keep the main factions apart so that negotiations for peace could take place and to allow the aid agencies to operate without the risk of their workers being killed.

American Marines landed near Mogadishu in December 1992, in an operation reminiscent of those at Grenada and Panama. They were ridiculed for their D-Day-like landing on a humanitarian mission but the ridicule was not justified. It was more dangerous to venture into Somalia than into Grenada or Panama, even if the plan was to keep the peace and not to be aggressive. The Americans and their allies found the fighters of Somalia, regardless of their political affiliation, dangerous and unstable. The most hostility came from those who followed General Muhammad Aidid. Aidid was declared a wanted man but several traps laid for him proved unsuccessful. Attempts to kill him from helicopter gunships were also fruitless, while the unintentional killing of Somali civilians from the air caused outrage. One helicopter crashed and its pilot, Michael Durrant, was captured. His terrified face on television turned many Americans against the whole US involvement.

In June 1993, followers of General Aidid killed 24 Pakistani soldiers and on 3 October 18 Americans were killed and others wounded during a fire fight. Under great pressure to pull out of Somalia, President Clinton ruled out immediate withdrawal, thus demonstrating that the US would not cut and run if some of its soldiers were killed. He set 31 March 1994 as the date for withdrawal.

The American decision was a political victory for General Aidid, who called a press conference at which he mockingly congratulated the US on having 'decided to address its past mistakes' - by which he meant its attempt to take him prisoner. In fact the point of the US policy shift was to call off the hunt for Aidid, which was widely blamed for converting a humanitarian mission into a war.

The need now was to concentrate on a political settlement that would prevent the country from falling apart after the US troops departed. Clinton's special envoy, the able Robert Oakley, met five of Aidid's lieutenants; as a result Michael Durrant and a wounded Nigerian, Umar Shantili, were released.

Oakley was also successful in persuading neighbouring African states, notably Ethiopia and Eritrea, to join in the peace-making. In addition, he helped organise an African mission to investigate and establish responsibility for the shooting of the Pakistani soldiers, though the UN and the US had no doubt it was Aidid. This move by Oakley had an important objective; it would enable the UN and US to negotiate with Aidid without officially ignoring a UN Security Council resolution calling for the arrest, trial and punishment of those responsible for the outrage.

The Americans insisted that Operation Quickdraw (withdrawal from Somalia) was not a retreat; nor should it be seen in that light. It was rectification of an error of judgement. Exit Day, or E-Day was actually a three-day operation, ending on 31 March. Most of the US troops were taken out by C5 Galaxy transport aircraft from Mogadishu airport. As 15 helicopters lifted off in formation, Somalis began looting office chairs, benches and electric fans from the hangar where the last Americans had assembled. The US land force commander, Major General Thomas Montgomery, announced that 11th-hour peace declarations, signed by the two chief warlords under pressure from the UN, would save Somalia from continuing anarchy. Few observers were so sanguine about the country's future.

Foreign intervention had brought aid to Somalia and saved thousands of lives, though no official expression of gratitude was made. During this unfortunate episode the Australian battalion serving in Somalia, under adverse circumstances, performed well and showed disciplined restraint. Only one Australian life was lost and that as a result of an accident within his own lines.

References

1. CIA intelligence on the looming disaster in Ethiopia was timely and thorough. The Americans were able to exert influence in the country long before Mengistu actually resigned and fled. US diplomacy, though criticised in Ethiopia, was responsible for the relatively peaceful transition of power which followed Mengistu's downfall and it averted further bloody civil war.
2. Ethiopia needs a defensive army and, after its losses in the final years of the war, it must restructure. The country is bankrupt and whichever arms-producing country is prepared to offer the government unlimited credit for the purchase of new arms will gain great political influence.
3. Most diplomats in Ethiopia and many officials in the foreign ministries of European countries have confidence in the EPLF's remaining stable and disciplined. Few nationalist movements anywhere have so much political goodwill.
4. Barre was a smuggler on a large scale; he also sent poaching teams into Kenya for elephant tusks and big-game skins. This trafficking helped to give him funds which he deposited abroad, ready for an emergency.

5. Somalia under Barre had the worst human rights record in the world, according to aid agencies.
6. *New Africa* magazine intercepted a letter from Morgan to Barre in which he suggested that the solution to 'the northern problem' was genocide of the Isaqs.

War Annual Nos 1 — 5 carried lengthy details of the progress of the campaigns, offensives and battles of the Horn of Africa.

India — Pakistan War

INDIAN ARMY'S TERROR CAMPAIGN
Background Summary

The Indian and Pakistan armies have been in a state of war over Kashmir since 1947 and the fact that the conflict has not spilled over into a more general war is due partly to international pressure and partly to the fear of both countries that such a conflict would be catastrophic. India holds two-thirds of Kashmir. Pakistan, with the other third, claims the entire region on the grounds that it is inhabited by Muslims. The Pakistanis say that Kashmir should have been given to them on partition in 1947.[1]

Pakistan has always been worried that, as India dominates the 800-mile Karakorah highway from Peking, it could cut it and thus separate Pakistan from its ally, China. The rivers that flow through Pakistani Kashmir have their source in India and this is another factor in the dispute.

UN observers — 36 of them in 1991 — have been stationed in Kashmir since 1949 but their presence does not prevent fighting. The observers merely ensure that it is reported to the UN. Strongly-held views in Islamabad and New Delhi prevent any compromise from being reached. The Line of Control — as the ceasefire line has been known since 1972 — is about 500 miles in length. Its highest point lies in the Karakoram Mountains in the north.

In 1990 the war intensified to dangerous levels when Indian troops killed civilians on a suspension bridge and Pakistani raiders planted their national flag in Indian territory.

The hostilities along the border between Indian Kashmir and Pakistani Kashmir are incidental to a greater struggle being waged inside Kashmir. Brutal and bloody, this takes place in the great Kashmir valley and especially in Srinigar, the summer capital. Muslim Kashmiris demonstrate and fight for Kashmir to be wholly governed by Pakistan, while other militia, both Muslim and Hindu, want total independence for Kashmir. Even within this basic division there are other factions among the 40 militant groups. The struggle became even more violent in January 1990 as the Indian Central Reserve Police Force and army units tried to control the subversive Muslim Kashmiris. By September 1990 the Indian army had 350,000 men in Kashmir.

As the uprising against Indian rule intensified, many people fled to refugee camps in Jammu, while others pushed through deep snow and mountain passes to reach the sanctuary of Azad Kashmir, the slice of Kashmir under Pakistan's control. India finally clamped down on the exodus and some hundreds of escapees were killed.

As India tightened its security in Kashmir, human rights organisations and international observers were alarmed by the increasing evidence of rape and torture of Muslim women by Indian troops. A Delhi-based human rights group, the Committee for Initiative on Kashmir, published a detailed account of such atrocities in the valley, part of a campaign of terror.

The Soviet Union (when it still *was* the Soviet Union), the United States and Japan have put intense pressure on India and Pakistan to seek peace, but without success. For the Indians, at stake is Mahatma Gandhi's belief that the only way to govern nearly 1,000 million people of numerous faiths is through secular democracy. If Kashmir were allowed to break away over religious differences it could lead to a bloody Hindu backlash against the 100 million Muslims living in other parts of India. Neither India nor Pakistan has a strong political peace lobby. Any such voice that might once have existed has long since been drowned out by Right-wing religious parties on both sides calling for a greater war.

The War in 1991

The Jammu and Kashmir Front (JKF) started the war in the Kashmir Valley in 1988 but in 1991 it was being pushed aside by the well-funded heavily-armed Muslim fundamentalist groups. The JKF, which is a secular organisation, came under increasing pressure from the Pakistan government, which favours the extremist fundamentalists. In addition, the State Government which administers the Pakistan side of Kashmir contemplated banning the JKF because its independent stance was embarrassing the authorities.

The JKF wants an independent Kashmir free of both Pakistan and India. Amanullah Khan, its leader, acknowledged in November 1991 that there was danger of civil war between different groups. He understated the position. It is more likely that genocidal Afghanistan-style battles for supremacy will take place in Kashmir.

Muzaffarabad, capital of Azad (free) Pakistan, tolerates a JKF office with guards around it but the relationship was an increasingly tense one late in 1991. This reflects Pakistan's desire not to allow the independence movement to take root on its side of the border. Pakistan's support for the fundamentalists has changed the nature of the conflict in the Kashmir Valley and led directly to the decline of the JKF, once by far the largest armed group.

Guerrillas on the Offensive

More than 100 guerrilla organisations operated in the Valley at the beginning of 1992. Most groups favour unity with Pakistan, which all foreign observers say is clearly not the desire of most Kashmiri Muslims. Pakistan was fanning the flames of war.

The political instability which followed the assassination of Rajiv Gandhi came as an opportunity for the Kashmiri militants. Khalid ul Islam, chief spokesman for the Kashmir People's Liberation League (KPLL), one of the strongest groups, told foreign journalists:'The death of Gandhi will increase the chaos in India and the struggle for freedom in Kashmir. It is a golden opportunity. India must now accept reality or face the collapse of its state system.'

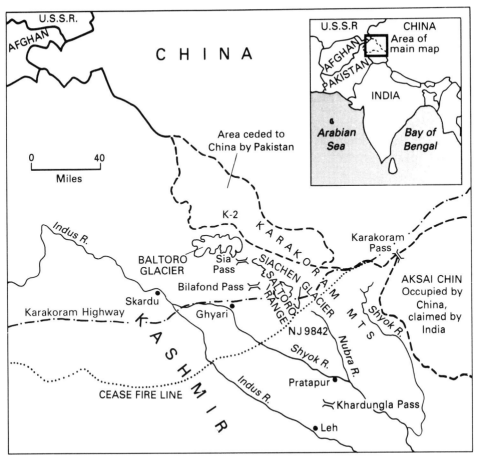

India-Pakistan Confrontation

	INDIA	PAKISTAN
Troops	1,260,000	520,000
Paramilitary troops	672,000	164,000
Tanks	3,150	1,750
Artillery	5,320	2,500
Aircraft	836	451

Gandhi's death was in every way an important moment for Pakistan in its 40-year war with India. The Nehru dynasty, which had been seen as consistently hostile towards Pakistan, was apparently at an end and the elections threw up a weak and unstable government in New Delhi. It was occupied with controlling separatist movements and Hindu fundamentalism and as a result had little appetite for embroilment with Pakistan over Kashmir. 'Pakistan is ready for peace', a Pakistani minister said, 'but the Indians have yet to respond positively.'

Since Kashmiris took up arms against India in December 1989, at least 3,000 of them have been killed and thousands more have been jailed and tortured. The militants claim that 3,000 young Kashmiri men are missing while more than 100 women have committed suicide when facing rape by Indian soldiers.

In mid-1991, militant leaders said that they were moving from a 'purely defensive guerrilla war' to the offensive in order to protect the population. They would continue to kidnap foreigners in Kashmir and hold them in 'safe protective custody' to protest against the lack of international concern over human rights violations in Kashmir.

The militants were heartened when the kidnapping of two Swedish engineers in Pakistan by the Muslim Janbaz Force resulted in the UN Secretary General making a direct appeal for their release and several human rights organisations condemning Indian atrocities in Kashmir.[2]

The summer of 1991 brought the war back to the border. In the first week of May the town of Atmuqam, on the River Neelam, and other villages on the Line of Control came under Indian shellfire. The UN Observers reported that 20,000 shells fell in the area in six days, killing 31 civilians and wounding more than 100. Great damage was caused and thousands of people fled from the region.

Pakistani heavy artillery retaliated and destroyed an Indian post. The two sides were so close across the river that each could identify with the naked eye the units opposing them. According to the Pakistani officers, the Indian aim was to terrorise villagers and cut the Neelam Valley road that provides the only access to the north of Pakistan Kashmir. The Indians for their part claim that Atmuqam is a centre for Kashmiri militants who, after being armed by Pakistan, cross the border to attack Indian troops. Pakistan denied this charge. Later in May India warned Pakistan not to undertake any 'misadventure in Kashmir'.

The tensions on the ground, where the two armies face each other are the reality; the diplomatic rhetoric about peace is a smokescreen to impress international opinion.

Separation in India

India experienced tremendous turmoil at the end of 1990 and into 1991. Experienced observers assessed it as the worst since the bloody days of partition in 1947. Apart from the fighting in Kashmir (described above) there was violent conflict in the Punjab between Sikh insurgents and security forces. The death toll for 1990 was 3,437 and likely to be even higher for 1991. The army was sent in to break the Sikh guerrillas' hold over the farming country in the border districts adjoining Pakistan.

There was only slightly less violence in the north-eastern state of Assam, as army units frequently hunted through jungle terrain for secessionist guerrillas who have paralysed the state. During 1990, 500 people died in Hindu–Muslim clashes, notably in Hyderabad — capital of Andhra Pradesh — and several cities in Uttar Pradesh and Gujarat. The fighting dwindled only after the army took control of the streets from the ineffectual and often anti-Muslim police units. The death toll for 1991 was 750.

Against the background of separatist coups and attempted coups elsewhere in the world, particularly in the Soviet Union and Eastern Europe, Indian political observers had begun to wonder, in 1991, whether the union could survive. No authoritative voice was actually warning of fragmentation, if only because the democratic system is still resilient enough to accommodate dissent. But India is veering in an ever more authoritarian direction and, as the pressures on its institutions increase, the response from New Delhi is coercive. The central government's dilemma is that neither a tough nor an accommodating approach promises a solution. Jammu and Kashmir, Punjab and Assam complain about a legacy of broken promises and subterfuge from Delhi.

Other influences are widening the gap between the government and its officials in New Delhi and regional insurgents. A major one is support from Pakistan, which foments rebellion in Kashmir and Punjab. Another problem is the rise of the Hindu fundamentalist movement and its campaign to make India into a Hindu *rashtra*, an avowedly Hindu state in which the god Rama would serve as a national symbol. This ambition terrifies India's 110 million Muslims. It also alarms Sikhs, who learned about Hindu ferocity when 2,500 Sikhs were killed in rioting in New Delhi in 1984.[3]

The Great March

In February 1992, the JKLF leader, Amanullah Khan, organised a march of Pakistani Kashmiris to express solidarity with a Muslim revolt in Indian Kashmir. The intention was to cross the Indian frontier at Chakothi and other points, to link with the Indian Kashmiris and plant flags on the Indian side of the India–Pakistan ceasefire line.

The Pakistan army, under instructions from the government, had orders to stop the human waves from reaching the frontier. The government feared that breaches of the frontier might lead to retaliation by the Indian army, which in turn could cause another war between the two countries.

Despite all attempts to halt them, with rolls of barbed wire and tear gas, the massed Kashmiris pushed out of Muzaffarabad, capital of the Pakistani-controlled part of the state. Pakistani military engineers blocked the road by dynamiting rock overhangs and bulldozing hills of earth across the tracks. When the Kashmiris surmounted these obstacles the authorities placed 400 policemen at the narrow bridge across the Jhelum River and near Chinari. Some shooting took place here.

Amanullah Khan called off his great march on 14 February when army commanders told him that anybody who tried to cross the final barrier would

Conflict in India

be shot. The militants, wet and bedraggled after 24 hours of torrential rain and snow, struggled back to Muzaffarabad, 30 miles away.

Late in October 1993 Benazir Bhutto made a comeback as Pakistan's prime minister and within a week she had a crisis on her hands. Thousands of Indian troops had surrounded the Islamic shrine of Hazratbal, causing angry protests and demonstrations among Muslims in Kashmir as well as in Pakistan.

The shrine is the holiest in Kashmir but Indian officials claimed that Kashmiri separatists had been concealing weapons in the mosque with the intention of turning it into a fortress. They demanded that the militants surrender with their arms. In the meantime they enforced a curfew. In Srinigar and other Kashmiri cities large demonstrations occurred and Indian troops opened fire, killing about 100 people.

India accused Pakistani intelligence agents of fomenting trouble in Kashmir - a charge accepted by foreign observers - and expelled four Pakistani diplomats from New Delhi. Pakistan denied the allegation and expelled four Indians from Islamabad. India's next move was vastly to strengthen its forces on the Pakistani border. At the same time India offered the militants in Hazratbal safe passage if they would vacate the mosque peacefully and leave their arms behind. The hundred or so fighters refused, saying that they would blow up the mosque - 'we prefer martyrdom to surrender'. The Indians cut the water, electricity and food supply.

The Indian government of Narasimha Rao dreaded gunfire in and around Hazratbal, knowing that the Muslims would portray it as an assault on the mosque by Indian infidels. India's own 120 million Muslims would then riot; they are still deeply angry over the destruction of the Ayodhya mosque in Uttar Pradesh by Hindu militants in December 1992.

Bhutto and Rao do not want conflict between their nations and between them they managed to defuse the Hazratbal crisis. However, there is little possibility of the greater problems over Kashmir as a whole being resolved.

References

1. A 1948 UN Resolution called for a plebiscite in Kashmir to choose between joining India or Pakistan.
2. One of the most damning reports was written by a Mr. S. M. Yassin, the deputy commissioner of Kupwara town and district, and an official of the Indian government. Mr. Yassin visited the town of Kunan Poshpura, investigated accounts of an appalling night of rape, and prepared a report which is unequivocal in its condemnation of the army's behaviour. He describes the soldiers as behaving like 'violent beasts'. He was rebuked for his outspokenness but the government quickly ordered an inquiry. However, it was made by the army itself, which meant nothing. Foreign travellers in the region confirm the Yassin report. British journalists Anthony Wood and Ron McCullagh reported the Kunan Poshpura incident for *The Independent*, London.
3. Simranjeet Sign Mann, leader of the *Akali Dal*, a militant Sikh party that demands independence for Khalistan, the militants' name for an independent Punjab, said: 'Anyone who rules in New Delhi must have the Hindu vote and Hindus have become the most intolerant of the minorities.' *Time Magazine*, 24 December 1990.

Iraq War (Or Second Gulf War)

THE PROFIT AND LOSS ACCOUNT

The war between Iraq and the United Nations (or the War Coalition) was described in detail in *War Annual No. 5*. Certain aspects of the conflict have become apparent since the fighting ended and are dealt with in this volume. In addition, it is necessary to reflect on the military significance of the war, though its more profound lessons may not be apparent for years. I prefer to call the conflict the Iraq War to distinguish it from the Gulf War, the eight-year war between Iraq and Iran, which was progressively described in *War Annuals 1 — 5*.

Course of the War in Brief

Britain gave Kuwait its independence in 1961 and almost at once Iraq attempted to take over the territory by force in pursuance of a claim that it was actually part of Iraq. When the attack was repulsed and an Arab peace-keeping force organised to ensure Kuwait's integrity, Iraq recognised its sovereignty (1963). Disputes over borders and oilfield ownership continued and Iraq, under Saddam Hussein, was ever more aggressive. In mid-1990 Kuwait was so apprehensive that on 20 July it placed its armed forces on alert, as did Saudi Arabia two days later. Iraq invaded on the night of 2 August. On 6 August the UN Security Council endorsed economic sanctions against Iraq and a naval blockade was imposed.

The US army began to deploy forces in Saudi Arabia on 7 August and British units soon followed. In a five-month period, a massive defending force was built up, with contingents from numerous Western and Arab nations. The intention was to drive the Iraqi forces from Kuwait if Saddam Hussein did not voluntarily withdraw them. In the meantime many negotiators tried to bring about a peaceful settlement. The Security Council set a deadline for 15 January 1991, and when Iraq made no positive response a strategic air campaign began on 17 January. It had two objectives — to destroy Iraq's capacity to wage war and to cut off the Iraqi forces in Kuwait from Iraq itself. Despite enormous destruction to its infrastructure, Iraq proved remarkably stubborn. The UN forces — which were overwhelmingly American — commenced a tactical air war to soften up Iraqi defensive positions in readiness for a ground assault, which began on 24 February. On 27 February Kuwait City was recaptured and the following day President Bush stopped the offensive, and a formal ceasefire was negotiated. It brought about a stand-off between Iraq and the UN, but further violent conflict soon erupted within Iraq. It has been calculated that one Allied serviceman was killed for every 700 Iraqis.[1]

Military Factors in the Allied Victory

The Allies had so many superior assets at the beginning of the conflict in August 1990 and developed so many more during the course of the war that it is pointless to try to list them in order of 'importance'. They were:

- Overwhelming air power in sheer numbers and variety of aircraft; the skill of the crews and the availability of sophisticated ammunition.
- Real-time signals intelligence (signint) from several sources, including satellites, and its rapid evaluation by computer.
- Superior technology including, most importantly, that used in the development and deployment of precision weapons.
- Massive naval forces in the Gulf. They imposed the UN blockade and were able to fire missiles against distant Iraqi targets. Allied sea power also indicated an ability to mount large-scale amphibious landings.
- Logistical support in volume and organisation not seen since the Second World War.
- Professional leadership which brought to the Allied operations a high degree of co-ordination and control far in excess of that available to the Iraqi armed forces.
- Large numbers of well-trained personnel.

Factors Operating Against the Allied Military Force

It has been fashionable to extol the many advantages enjoyed by the US-led coalition. Contrary influences have been given less exposure, but they should be recognised even if they are not necessarily direct military factors. The war was fought and won despite frictions which were at times severe. The bureaucratic web and the religious leaders in Saudi Arabia were particularly troublesome to the Western leaders, both those in the distant capitals and those on the spot, such as General Schwarzkopf.

The military leadership found the Saudi officials always difficult to deal with and at times deliberately obstructive. Even though the Western troops were defending Saudi territory, the Saudis were unco-operative about water supplies, use of roads and ports and provision of storage facilities. When the obstruction reached a critical level the Western Allies, and in particular the Americans, bypassed the Saudi bureaucracy or ignored it. At no time were the Western forces popularly received and there were numerous cases of troops being insulted and even assaulted.

The Allied leaders were handicapped by the political need to involve Arab troops in some way in the campaign against the Iraqi forces. To a limited extent the Saudi forces were incorporated and a few of their pilots flew operational sorties. There was no great trust in the Egyptians and even less in the Syrians and Moroccans. The use of Arab troops in forward positions confronting the Iraqis was a good one. Even more shrewd was the intention to move the trusted Western troops — notably the Americans, British and French — through the Arab screen when the time came to attack the Iraqi forces. In this way no offence was given to the Allied Arab armies since they could claim to have been in the front line. They contributed little to the actual assault.²

Perhaps the great problems faced by the Western governments was opposition to the war at home and the fear that casualties would become politically and socially intolerable. President Bush and his administration feared casualties on the scale of those suffered in Vietnam (where 55,000 Americans were killed). This apprehension among the politicians inhibited the Generals' freedom of action. They realised that they had to win quickly and decisively before their political masters called a halt to the war.

Western spokesmen presented the war as one of self-defence; the use of force against an aggressor was permitted under the UN Charter. Over and again the Allied leaders stressed that they were engaged in a war to drive Iraq out of Kuwait, *not* waging a war to defeat Iraq. Some of them may actually have believed in this distinction from the outset but in any case they came to believe it through sheer repetition. Of course Iraq had to be defeated, if only in order to expel its armies from Kuwait. However, the repeated assertion that the Allied objective was merely to drive Iraq from Kuwait was fundamental in maintaining international support. President Bush went to great lengths to emphasise the 'limited objective'.

In the end, the President came to believe his own protestations about a limited objective and he brought the conflict to an end before Saddam Hussein's potential to wage war again had been destroyed.

The Media and the War

No war has been so comprehensively reported. The media had almost total access to the military build-up and, as a result, Saddam Hussein and his Generals must have had a thorough knowledge of the various troop formations, weapons, equipment, armour and ammunition about to be used against them. Similarly, once the air war began virtually every aspect of the operations was open to television scrutiny: even the fine technical points of bombing were analysed on screen. During press briefings at Command HQ in Riyadh, military spokesmen gave lengthy, detailed descriptions of operations and their statements and associated maps and diagrams were transmitted worldwide.

Throughout the war numerous 'enemy' journalists operated from within Baghdad and, while they were not free to say exactly what they might have wished, their reports contained much information that was useful to the intelligence services.

Despite the extraordinary freedom given to the media by the Allied high command, many complaints were made about censorship and manipulation. Some reporters and their employers protested that the military spokesmen had deceived them in not providing accurate information about the land offensive. This raises the question of just how much information the media can legitimately expect to be given. Security considerations must be paramount, even if this requires deception of the media.

Meticulous Campaign Timetable

Western spokesmen and their propagandists made great play of the claim the

the Coalition against Saddam had fought a clean war, one academic writer referring to the Allies as having fought the war in 'a reasonably clean and legal fashion'.[3] 'Clean' apparently meant that the Allies did not use chemical, nuclear or biological weapons or napalm. Saddam Hussein, in contrast, while prepared to use all three, in the event used none of them.

Throughout the war the media, echoing the military, gave the impression that a precision war was being fought against military targets, not against people. The term 'body counts', so prevalent in the Vietnam war, did not appear in official reports or press stories. Certainly, during the earlier part of the war most attacks were made against clearly-defined military targets, such as munition dumps, missile sites, command posts and barracks. Before long 'dual-use targets' were specified as legitimate. This category included power stations, bridges and telephone exchanges.

Keeping to a meticulously planned programme, the Allies moved on to the destruction of Iraqi forces in Kuwait and south-eastern Iraq. Weapons used included area-impact munitions which had been developed as a result of Vietnam experience. The principal types employed were cluster bombs, the Multiple-Launch Rocket System (MLRS), slurry bombs and fuel-air explosives. An immense number of casualties were caused by the cluster bombs, notably the Rockeye 11. This 500lb bomb scatters 247 grenade-size bomblets which in turn explode into something like half a million lethal fragments or shards. The US Air Force and Navy together dropped 34,000 Rockeyes. The Royal Air Force Jaguar aircraft dropped thousands of the British-made BL755 cluster bomb.

In the Iraq War, the American and British armies used the MLRS for the first time in combat. A tracked vehicle carries 12 rockets, all of which can be fired within a minute. With a range of 20 miles, the missiles are generally set to detonate over a target area of 60 acres. A single salvo scatters 8,000 fragmentation grenades with devastating effect. During the final two weeks of the war the US and British forces fired 11,000 missiles between them, in addition to many thousands of other anti-personnel projectiles.

Just before the end of the war, Iraqis fleeing from Kuwait along the Basra road were killed in large numbers by aircraft which strafed the columns of trucks, cars and armoured vehicles. Many of the victims were burned to death and some were vapourised. When films and reports of the slaughter were seen in the West, there was an outcry about the 'massacre'.

Further protests were made when, in August 1991, newspapers reported that Iraqi troops had been buried alive during the US ground attack. The first impression was that these events had taken place at the height of battle. In fact, the action had been rehearsed as a way of minimising American casualties.

Abrams tanks fitted with bulldozer blades drove parallel to the Iraqi trenches, filling them in and burying Iraqi soldiers alive. There is no doubt that the Iraqis were firing their weapons to the last moment and that they were not surrendering. Many people in the West were horrified by the tactic of burying enemy soldiers alive but probably most of those who objected still nurtured the idea that there could be a clean war. There can be no such thing.[4]

The object of any army at war, whether in an invading or defensive role, is to win quickly with a minimum of casualties to itself. To achieve this desirable

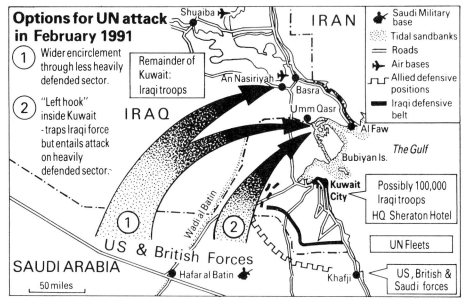

The Iraq War

end, it is sometimes necessary to inflict heaby casualties on the enemy. Saddam Hussein was prepared to inflict massive casualties on his enemies, as he demonstrated during his war against Iran. Had he or his Generals thought of using bulldozer tanks to destroy enemy trenches and bury their occupants, can there be any illusion that they would have declined to use this tactic on the grounds that it was dirty? They had already used chemical weapons against their own Kurdish people.

The simple fact is that all battlefield deaths and wounds are horrible and to suppose that it is possible to fight a clean war displays not only ignorance but lack of common sense. It is worth noting that the carnage on the road to Basra and the reports of Iraqi soldiers being buried alive aroused greater anger in the West than reports from UN personnel in Iraq who describe a great post-war increase in ill-health and in infant mortality, as a direct result of the destruction of the Iraqi civil infrastructure, such as the water supply system.

War Casualties

While Allied casualties are known precisely, those for Iraq vary wildly, according to the methods used to assess them and the source which has supplied them. The US forces lost 389 killed, some of them accidentally, and 457 wounded. Other allied forces suffered 77 dead and 830 wounded.

In an attempt to find a figure for Iraq's military casualties, the US Natural Resources Defence Council, an environmental group, filed a request under the Freedom of Information Act. The Defence Intelligence Agency responded in June 1991 by releasing an estimate of 100,000 Iraqi soldiers killed and 300,000 wounded. But, said the DIA, these figures contained an 'error factor' of 50 per cent or higher. In short, the statistics were worthless.

Pentagon sources generally have not sought to publish refined figures if only because figures produced during the Vietnam War were greeted by some observers as preposterously high or as callous boasting. British Ministry of Defence officials have estimated Iraqi losses as 30,000 dead and 100,000 wounded. Some UN officials who have had opportunities to travel widely in Iraq say that they have not seen enough injured veterans to justify a wounded-in-action estimate of anywhere near 300,000.

All figures are based on a series of extrapolations. This is the basic analysis: (a) calculate the approximate number of enemy troops on service at the beginning of the war; (b) subtract the number of prisoners and the estimated number of deserters; (c) apply to the remaining total certain standard ratios — for each 10 soldiers engaged a specific number can be counted as dead and a certain number wounded.

The starting figures come from satellite and aerial-reconnaissance photos, interrogation of POWs, reports from special forces operating behind enemy lines, from spies, and from historical ratios of percentage of casualties in past wars. Few actual counts of bodies were made during the Iraq War. Most dead Iraqis were hurriedly buried by their comrades before the onset of the land war or by Saudi burial parties afterwards and generally no tally was kept.

Because the war was largely fought by air attacks against ground targets, Allied officers have tried to calculate the casualties from the numbers of tanks,

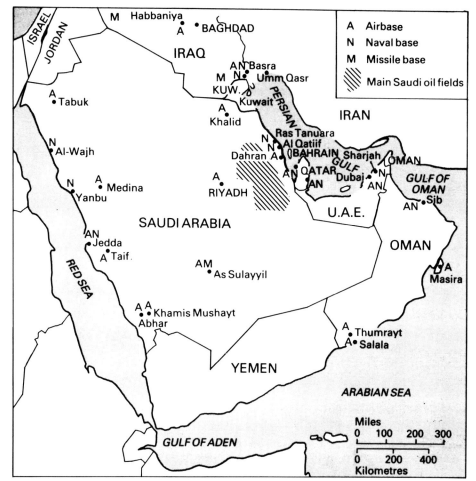

The Iraq War

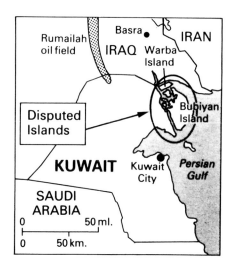

other vehicles and artillery pieces destroyed. But this is fallible as aerial photographs produce no evidence of how many men a wrecked APC was carrying, how many were killed or wounded and how many escaped unharmed.

The US figure for enemy soldiers in Kuwait and southern Iraq when the war began was 540,000. This is much too high. It was based on satellite pictures from which allied commanders counted the numbers of divisions deployed. But we now know that many units were well below strength. According to information derived from prisoner interviews, the desertion rate was high — in fact, even higher than the 150,000 the Pentagon analysts estimated. At the start of the land war Allied troops found the enemy defences thinly manned. The assumption must be that there may not have been enough Iraqis on hand to suffer 300,000 casualties (the DIA estimate), even if every last one had been killed or wounded.[5]

Casualties From Friendly Fire

The rate of 'fratricidal casualties' — or deaths and woundings from own-fire — during *Operation Desert Storm* was ten times as high among US troops as in any battle or campaign of the century. In August 1991 the Pentagon confirmed that 35 of the 146 Americans killed in action and 72 of the 457 wounded were victims of 'friendly fire'. US fire also destroyed seven MK1A1 tanks and 20 of the 24 Bradley Fighting Vehicles lost in battle. US artillery even fired on the battleship *Missouri*, although the small Iraqi navy had nothing larger than corvettes. In addition, American A-10 tankbuster aircraft killed nine British soldiers.

US Marine Corps Lieutenant General Martin Brandtner, who revealed the figures,[6] conceded that the number of friendly-fire incidents was abnormal. 'These incidents were due to a combination of featureless desert terrain, large, complex and fast-moving formations, fighting in rain, darkness and low visibility, and the ability to engage targets from long range', he said.

Certainly visibility was a crucial factor. Nocturnal encounters were frequent and during heavily overcast days the units relied on infra-red heat-sensitive imaging devices which, at great distances, provide only a fuzzy image. It might be imagined that a military system that can produce the Patriot anti-missile-missile and direct bombs through a ventilation shaft might also find a foolproof way of distinguishing friend from foe. However, the M1A1 tank's 120mm cannon, though lethal at 3,200 metres, could not positively identify targets at that range through its infra-red optics.

All US forces have failed to solve the Identification Friend-or-Foe (IFF) problem, despite an IFF development programme that has been in progress at Wright-Patterson Air Force Base in Ohio since 1980. There is still nothing more sophisticated than an identification panel taped to friendly vehicles and small transmitter beacons which do not always work.

The US strike against Warrior APCs of the Royal Fusiliers was particularly blameworthy. The Warriors were in formation, at rest in the mid-afternoon on a day of perfect visibility, when they were attacked by two A-10 aircraft.

Having flown over the column, the A-10 pilots came back to fire their infra-red Maverick missiles, which killed all the troops in the APCs.

The A-10 pilots had been looking for Iraqi Russian-made T54/55 tanks, which bear no resemblance to a Warrior APC. The enemy armour was $2\frac{1}{2}$ miles distant. During the same phase of battle, two US reconnaissance aircraft, though flying 5,000 feet higher than the A-10s, correctly identified the Warriors. The reconnaissance pilots saw the identification panels, an inverted V carried by all allied vehicles in the Gulf. These signs were clearly visible on thermal identifying equipment and to the naked eye. One of the A-10s made two passes before coming in to make its attack and its pilot should have been easily able to recognise the Warriors for what they were. Following a USAF inquiry, the pilots were cleared of blame.

A British Board of Inquiry also reported on the episode and stated that it was unable 'to establish why the vehicles were misidentified by the A-10 pilots'.[7] This report confirmed that all markings were correct and in place and called for a study to find an air recognition system that would 'help to prevent such tragedies in the future'. The US government refused to allow the American pilots concerned in the incident to appear at the British inquest into the soldiers' deaths.

It should be noted for the record that seven US combat engineers were killed while stacking cluster bombs, an accident which raises questions about safety procedures.

The Pentagon Analysis

Between the end of the war and August 1991, Pentagon officials[8] worked on an 'Iraq War Report', the first version of which ran to 178 printed pages. Some of the report so disturbed senior military bureaucrats that they attempted to blunt the direct and implied criticisms about the handling of the war. Even now some parts of the report are unlikely to be seen except by senior officers and officials. Enough is known, however, to indicate that the US war machine had numerous imperfections when it went to war against Iraq.

The US forces were ill-equipped to face biological attack, though this was considered a major risk, and they had fewer than half the hospital beds needed. The High Command was over-dependent on satellite links, which were vulnerable to jamming, and it was short of tactical intelligence. Mine-clearing equipment was insufficient in quantity and inadequate in quality and tank ammunition was in short supply.

The poor quality of tactical intelligence was well known throughout the US armed forces. Some commanders have said that they wasted time and effort in attacking targets that had already been destroyed while others remained untouched because their position had been wrongly reported. Many assessments were vague, or so qualified that they were useless. 'My main intelligence problem was making sense of the intelligence I received', one Major General said.

The Defence Secretary, Dick Cheney, in an introduction to the report, states that the campaign was 'brilliantly orchestrated'. Most observers would agree with this as a generalisation. Cheney also says that victory was 'neither easy nor certain'. This phraseology could be interpreted as a way of stressing the professionalism of the armed forces. However, Cheney actually uses this language to admit obliquely that much was wrong with the US war effort.

For instance, despite intelligence reports (correct as it happened) that Iraq was prepared to use biological weapons such as anthrax, no detection systems were in place until the war was nearly over. For four months after US troops arrived in Saudi Arabia no vaccines were available.

Allied commanders were concerned, according to the report, that the Iraqi air force could have launched a single devastating air strike, which would have damaged public support for the war as badly as the Vietcong's disastrous Têt offensive of 1968. Such an attack would also have exposed the shortage of hospital beds — only 2,642 instead of the 7,350 which Central Command believed to be necessary.

One disclosure is startling in its implications. The Iraqis themselves could have used the powerful Global Positioning Satellite system, which allowed the US forces to navigate through featureless desert. The US was unable to encrypt the satellite's signals because their troops were dependent on inferior 10,000 watt commercially-available receivers, hurriedly acquired at the start of the crisis.

Among the hundreds of problems identified in the report some stand out, for instance, the slowness of industrial suppliers in responding to orders for food, ammunition and protective gear. Had the US been forced into an early ground war, the army would have been hard put to maintain the troops in the field. The mobilisation of reserve shipping took twice as long as required by law and, according to the report, was 'alarmingly protracted'.

While the report does not say so, its findings strongly suggest that had Saddam Hussein and his Generals been more professional in their conduct of the war the end result would indeed not have been 'easy or certain'.

The Ahtisaari Report

On 18 April 1991 the UN Under-Secretary-General, Marti Ahtisaari, reported to the Security Council following his exhaustive fact-finding tour of wartorn Iraq. His observations and assessments have direct military significance for they documented the degree of destruction caused by the Allied air offensive and the UN blockade with a thoroughness that has not been equalled since. His report received little notice from the world's press, largely because it was submerged by the great volume of news about the flight of the Kurds. These were among the main points, quoted directly from the report:

- Iraq is devastated. The conflict has wrought near-apocalyptic results upon the economic infrastructure of what had been, until January 1991, a rather highly urbanised and mechanised society. Now most means of modern life have been destroyed or rendered tenuous. Iraq has, for some time to come, been relegated to the pre-industrial age but with all the disabilities of post-industrial dependency on intensive use of energy and technology.

- Approximately 90 per cent of Iraqi workers have been reduced to inactivity.
- All electrically-operated installations have ceased to function, with devastating consequences for water supply, public health and food distribution.
- Without electricity or refined oil fuels, food that is imported cannot be preserved or distributed; water cannot be purified; sewage cannot be pumped away and cleansed; crops cannot be irrigated; medicaments cannot be conveyed.
- Food stocks are near exhaustion; before the war Iraq imported 70 per cent of its needs but as sanctions take their toll and this year's harvest is compromised by the general devastation, the Iraqi people may soon face a further catastrophe which could include epidemic and famine.

Germany, after the Allied bombings of 1942-45, was the previous model for infrastructural devastation caused by strategic saturation bombing. But that country was able to recover relatively quickly because of massive American aid and finance. Iraq can expect no such aid. North Vietnam, which was even more heavily bombed — that is, by weight of bombs — suffered less than Iraq because it was almost wholly an agrarian society without an industrial infrastructure and without modern amenities of life. It was able to absorb the punishment inflicted upon it. Iraq possessed much more that could be damaged than North Vietnam had. War analysts had long wondered to what extent a country might be damaged through aerial and long-range missile bombardment. The Ahtisaari Report on Iraq provides the answer.

Iraq's New Challenges

During the first half of 1992, Saddam Hussein's armed forces showed that he and they were prepared to risk further confrontation with the Western coalition powers. For instance, while making no major offensive against the Kurds, the army moved several strong units closer to Kurdish areas. There was also an increase in conflict between government and Kurdish *peshmerga* forces. In April, fighting was reported on the Erbil–Mosul highway and Iraqi army tanks were captured in Suleimaiyah.

In the last week of March the UN again demanded that Baghdad must produce a comprehensive plan for destroying the ballistic-missile capability. As a warning, the Pentagon announced that about 50 Iraqi nuclear, chemical and ballistic-missile facilities were targeted for air or missile strikes. The US aircraft carrier *America*, carrying 85 aircraft, was sent into the Gulf. Iraq appeared to be ready to comply but it was not possible to trust any statement made by Saddam's regime.

On 5 April, Iraqi aircraft shot down an Iranian warplane that had, with other bombers, attacked an Iranian rebel camp at Khalis, in Iraq. The bombed guerrillas belonged to the largest Iranian rebel group, the Mujahideen Khalq. Ever since the end of the Iran–Iraq war in 1988 Iraq has supported the Iranian rebels hostile to the Teheran regime.

Saddam Hussein used the incursion of Iranian warplanes as an excuse to move anti-aircraft missiles into Kurdistan, 20 miles north of Mosul, where the allies had previously forbidden Iraqi military movements. That is, Iraq had crossed the forbidden 36th parallel, which bisects Kurdistan. North of this line, American, British and French aircraft, based in Turkey and on an aircraft carrier in the Mediterranean, are constantly patrolling. Allied fighter-bombers fly low overhead when any skirmish between Iraqi and Kurdish fighters breaks out.

The Allied High Command responsible for 'Iraqi Watch' reported that the missile batteries and their associated radar were a threat to its aircraft. UN aircraft on patrol within the exclusion zone are equipped with missiles designed to home in on radar emissions and they would certainly be used should Saddam's forces become aggressive.

In April 1992, Saddam had more confidence in his military strength than at any time since the Gulf War. His new offensive to clear rebels from the marshes of southern Iraq was evidence of this. It was clear to foreign observers that the time must come when the UN, or the Allies acting on behalf of the UN, would be required to put their threats to attack Iraqi targets into effect in order to stop further aggression by Saddam.

Throughout 1993 there were rumours that Saddam Hussein was at risk of being overthrown by one or other of his many enemies but on each occasion he survived and even gained in strength. During 1994, as in the previous three years, Iraq's problems were related to its three ethnic areas: the Kurdish north, virtually an independent state and protected by Western air forces; the Sunni centre controlled by Saddam; and the Shia south which is open to Saddam's ground forces though his air force is excluded.

The southern Shia area, especially the marshes, were the major focus of international concern in 1993. As the year ended the UN was investigating reports that Saddam's forces had used poison gas in attacks against Shia villages and rebels in the marshes.

Saddam's administration is continuing to drain the marshes, mainly in order to facilitate access by the army. This vicious programme is condemned as a human rights and ecological catastrophe. Yet is it unlikely that southern Iraq will be declared a safe haven; the army is present in strength and could not be dislodged without a full-scale invasion.

Meanwhile, Saddam's survival ensures the continuation of UN sanctions. Resolution 687 stipulates that the oil embargo will remain in effect until Iraq has dismantled its long-range missiles and non-conventional weapons programmes. Resolution 715 prevents the lifting of sanctions until a long-term monitoring programme is in place to ensure that Saddam's programmes of mass destruction are not revived. The International Atomic Energy Agency (IAEA) and the UN Special Commission (Unscom) have yet to declare that Resolution 687 has been complied with.

Kenneth R. Timmerman, who conducted an investigation on behalf of the US House Committee on Foreign Affairs, states that Iraq has already rebuilt about 90 per cent of its weapon plants. According to reports from both IAEA and Unscom many of these plants have been 'converted' to making conventional weapons systems not covered by the UN resolutions. They include T-72 tanks, ballistic missiles and heavy artillery.

In 1994 it was distressingly clear that the political opposition in exile was no threat to Saddam; only the opposition within the country means anything. Economic conditions deteriorate year by year but the populace is too cowed to protest. Iraq continues its search for friends but even the once sympathetic countries, Jordan and Yemen, are increasingly distancing themselves. Saddam constantly tries to gain support by exaggerating the Iranian threat to all Arab countries.

Self-Inflicted Wounds

On 15 April 1994 two US F-15C fighters enforcing the no-fly zone over Iraq shot down two UH-60 Black Hawk helicopters, killing all 26 people aboard. In this terrible instance of 'friendly fire', 21 of the victims were army officers from the US, Britain, France and Turkey; five Kurdish passengers were also killed. The accident wiped out the senior staff of the Allied Military Co-ordination Centre. Among the dead were Colonel Gerald Thompson, in charge of the command, and his newly selected replacement, Colonel Richard Mulhern.

The fighter pilots mistook the helicopters for Iraqi aircraft violating the no-fly zone, though there had been no reported incidents over northern Iraq since January 1993. The helicopters had been picked up on radar by a US Air Force AWACS reconnaissance plane, which called in the fighters. They gave no warning to the helicopters to land.

It seemed possible at the time that the helicopter crews did not respond to the pilots' IFF (Identify, Friend or Foe) call. Some investigators said that the accident had more to do with human error than technical malfunction.

Lives lost to friendly fire are a devastating cost of battle and almost a quarter of the 148 American combat deaths in the Gulf War resulted from accidental assault by their own side. However, even during the war, when hundreds of planes from more than 24 nations filled the skies, none of these deaths involved aircraft firing on one another.

The accident happened at a time when Saddam Hussein had been making bellicose gestures and he quickly exploited the US disaster in his propaganda broadcasts. 'Allah himself', Saddam said, 'has struck down "agents of the evil American empire"'.

References

1. The calculation was made by Dr. Hugh Smith of the Australian Defence Studies Centre, in *The Military Significance of the Gulf War*, published by the Centre in July 1991.
2. Syria contributed 4,000 men and some armour in order to win favour from the Saudis and the Americans. This they achieved. Saudi Arabia paid the Syrian government 2 billion dollars as 'goodwill'. The Bush Administration changed its public posture towards Syria, even though Syria remained one of the two states (the other is Libya) which provide encouragement and financial support for terrorist organisations. Syrian co-operation with the US ended in March 1992 when political and press vilification of Saddam Hussein ended and criticism of the US resumed. In the same month, a consignment of Scud missiles made in North Korea reached Syria. Foreign diplomats in Damascus said that the end of Syrian attacks on Saddam Hussein probably indicated the belief of President Assad that Saddam would survive as leader of Iraq.
3. Dr Hugh Smith, *ibid*.

4. At what level of command the decision to use the burying tactic was made is unclear. My information comes from a colonel of a US armoured unit and, independently, from an Australian officer serving on secondment with a British unit. I understand that the primary intention was not to bury the enemy soldiers alive but to prevent them from firing their weapons. Some ran before they were engulfed. In July 1991 Iraqi burial parties, under American control, unearthed soldiers from trenches overrun during the Allied assault, according to Western correspondents on the spot. The Iraqi government appears to have made no formal complaint, perhaps because its army used the same tactic against Iranian troops in 1983, during the Iraq–Iran War. The development has not escaped the notice of various armies. A Pakistani officer told me that his army regards it as an innovation that might well be used to save lives in 'defensive wars.'
5. The Iraqis have issued no casualty figures but Jordanian sources with Iraqi military connections have told me that the total military casualties 'exceed 150,000.'
6. At a briefing in Washington on 20 August 1991.
7. The British details were disclosed in parliament by the Army Minister, Archie Hamilton, on 24 July 1991.
8. The report was published in June 1991 and a revised version was published in September 1991. Parts of both reports were circulated only on a 'need to know' basis.

Israel and the Palestinians

THE INTIFADA (ARABIC FOR 'SHAKING FREE')

Background Summary

The *intifada* or Palestinian uprising on the West Bank and in the Gaza Strip began in 1987. Palestinians in the territories administered by Israel believed that Israel would be unable to maintain large numbers of soldiers on active duty to suppress the disturbances. However, the Israeli security forces maintained their counter-pressure.

Military experts warned that the *intifada* would adversely affect Israel's combat readiness in a conventional war, but the introduction of new technologies in riot control, such as the 'gravel-thrower', led to a reduction in the number of armed patrols. This reduced Israeli casualties and provided less provocation for the Palestinians.

By 1990 more Palestinians were dying at the hands of their compatriots than by bullets from the Israeli army. The real battle was a Palestinian one for control of the *intifada*, with five contenders. They were Yasser Arafat's *Fatah*, the Marxist Democratic Front for the Liberation of Palestine, George Habash's Popular Front, the Islamic fundamentalist movement *Hamas*, and the middle-class families with strong ties to King Hussein of Jordan.

Previously united to some extent by a common hatred for Israel, these groups turned on one another with accusations of 'harbouring' traitors who collaborated with *Shin Bet*, Israel's internal secret service. Hundreds of victims were stabbed, hacked or battered to death during 1990 and, in effect, a Palestinian civil war was in progress.

Palestinian strategy against Israel was based on a form of economic warfare. This included the boycott of Israeli goods, strikes that kept Palestinian workers out of Israel and the creation of 'independent economic units' built around subsistence farming. These measures resulted in the loss of several hundred million dollars to the economy, which, however, readily readjusted to the Palestinian strategy. In particular, foreign workers were imported as a more reliable source of labour. Large-scale immigration of Jews from the Soviet Union meant that the Arab labour was not missed.

The *Intifada* in 1991

The Palestinians involved in the *intifada* suffered a serious blow to their morale with the defeat of Saddam Hussein, who had been seen as the Palestinian saviour. Yasser Arafat had repeatedly assured the Palestinians that Saddam was their champion, and they personally felt his humiliation at the hands of the Western powers. His barbaric treatment of the Kurds and the Shia

Muslims did not concern them. His defeat was much more important and significant. Arafat's support of the defeated Saddam does not appear to have weakened his reputation among the Palestinians of the Administered Territories, perhaps because Arafat has no rival for power.[1]

A new Israeli Chief-of-Staff, General Ehud Barak, and Major General Danny Rothschild, the defence ministry official in charge of the Civil Administration — the name given to the military government in the Administered Territories — brought about significant changes to the *intifada* in 1991. Barak took the first step on 21 June when he authorised the showing of a film which graphically documented the security forces' undercover units. It showed men of these units donning Arab disguises, including women's dress, to infiltrate Palestinian society to hunt down those on the wanted list and to shoot some of them.

It had been common knowledge since 1989 that *Shin Bet* agents had masqueraded as Palestinians and as foreign journalists and tourists. Among their prime targets were the killers of people suspected of being *Shin Bet* informers. These were members of Palestinian death squads which visited 'disloyal' villages to kill, to wound or to intimidate the people and force them to support *Hamas* or PLO factions.

The film might not have told the Palestinians much that they did not already know, but it astonished most Israelis. Some were horrified at the revelations, others were angry that 'state secrets' should be revealed in this way.[2] General Barak's main reason for breaching the traditional Israel secrecy about its undercover operations was psychological: he wanted the *intifada* fighters to know that they were not safe from Israeli countermeasures.

Two Right-wing cabinet ministers said that the admission of *Shin Bet* activity should never have been made. Rehavam Zeevi, a minister without portfolio, threatened to recommend that his party *Modolet* (Homeland) withdraw from the coalition which rules Israel. Rafael Eitan, a former army Chief-of-Staff, said: 'I cannot understand the stupidity of the decision. It damages Israeli intelligence and puts people's lives in danger.'

In response, Brigadier Nachman Shai, the chief army spokesman, said: 'The *intifada* activists should be aware of the fact that nothing is secure, that no one they can see around them can be identified as a friend or as an enemy.' He explained that trouble-makers would not be tolerated and, in one way or another, would be hunted down and dealt with. Immediately after the film's showing, police arrested 40 Palestinians suspected of involvement in the murder of 16 fellow Arabs.

The message was loud and clear and it seemed to get through. Almost simultaneously with the showing of the film, a leading activist in the *Fatah* group told a British journalist:

> Within a few weeks you are going to see a new trend in the *intifada*. We have learned from our mistakes. There is going to be a whole new method of resistance. The new policy will be more practical. There is going to be a ban on killings under any pretext. No more of these investigations and questionings of Palestinians by masked activists. There will be an emphasis on the political and non-violent aspects of the *intifada*. Popular committees

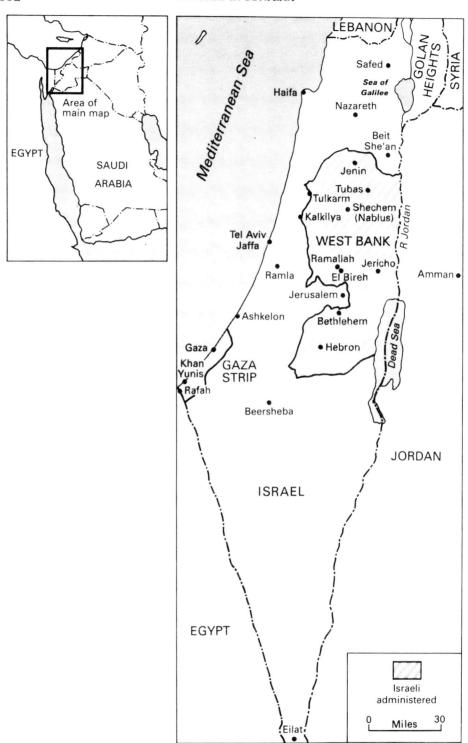

West Bank and Gaza: Areas of Intifada

will be established again. We shall adopt a tactic of dialogue with the Israelis themselves.[3]

But if this activist advocated a more subtle form of *intifada*, others were demanding even more violent action, throwing grenades rather than stones and causing explosions rather than setting fire to tyres.

Only two days after the showing of the controversial film, General Rothschild announced an easing of constraints on the Palestinians. This was a hint that Israel was prepared for economic co-operation with the Palestinians in return for an end of *intifada* activity. Together, the Barak-Rothschild messages stated in effect: 'Peaceful Palestinians have nothing to fear from the Israeli authorities; violent men and women will be dealt with mercilessly.'

For Israel, the timing of the initiative was right because Palestinians had been privately admitting their difficulties in running the *intifada*. Nothing was said in public because nobody wanted to draw Israeli attention to the organisational details of the insurrection. The Israelis were also aware that even after four years of the *intifada*, no clear leadership had emerged within the Administered Territories. The initiative of June 1991 was intended to prevent such a leadership from surfacing — unless it was prepared to co-operate with the Civil Administration.

The absence of a clearly defined leadership for the *intifada* is worrying prominent Palestinians, who anguish about the loss of authority over the teenage activists of the uprising. The 'strike forces' which at the start of the *intifada* had been encouraged to lead demonstrations and 'enforce discipline' are now out of control.

In 1991 the Black Panthers were the most militant of the groups of masked youths. Attached to the mainstream *Fatah* movement, the Black Panthers are discontented with the running of the *intifada* and are increasingly using their weapons against fellow Palestinians, despite appeals from the PLO and the Unified Leadership of the Uprising (ULU).

Not all the Black Panthers' victims are suspected of collaboration with the Israelis. Large numbers of people have been attacked for 'lax moral behaviour', personal retribution, family feuds or simply because they were a threat to some individual or group intent on controlling the streets. The extreme puritanism of the Islamic zealots of *Hamas* can be seen in the punishments meted out for lax moral behaviour, which might mean nothing more than being seen in a 'compromising position' with a woman.

Many people labelled as informers and killed were probably not informers at all. But people accused of collaboration with the Israelis have no defence and nobody will speak up for them for fear of suffering the same fate. Until recently Palestinians have avoided public debate on the subject of killing alleged collaborators. In June 1991, however, the journalist Khaled Abu Aker wrote in the weekly English edition of *Al Fajr*:

> By killing people without proper ruling or judgement from the correct legal institutions or people capable of carrying out such decisions, we are playing into the occupation forces' hands. At this fateful time which we are going through, I believe that it is the responsibility of our national

institutions and personalities to confront this dangerous occurrence, which has taken precedence over issues of the *intifada*.

Following the Gulf War, the Palestinians profess to have seen no sign of the new order promised by President Bush. Many once-active supporters of the *intifada* have wearied of the struggle and believe it needs to be overhauled. They are even willing to reappraise the Palestinians' outside representation. A vociferous minority has made many calls for greater democracy in the choice of their representative bodies, but there is little chance of this happening while Arafat remains in control of the PLO.

The opening of the peace talks in Madrid in October 1991 brought articulate, moderate Palestinians to the world's attention but since all the negotiators are PLO nominees and supporters they are unlikely to convince the Israelis of the sincerity of their protestations about peaceful co-existence.

The *intifada* is not dead. Palestinian stabbings of Israelis, including attacks in areas of Jerusalem that had been considered safe, occur frequently. The leaders of the *intifada* will find it difficult to establish order among their young militants. Their first task, however, is to reassess the strategic goals of the *intifada*, if it has not already outlived its purpose which was to bring the focus of the world attention onto the Palestinians and their plight.

The Black Panthers symbolise the continued spirit of the *intifada*. They sleep rough in the caves of the hills of the West Bank where an entire searching army would probably not find them. They rely on sympathisers for food, clothing and baths and they practically run the crowded cities of Nablus and Jenin, in the northern part of the West Bank. Almost every day the Panthers are responsible for shootings and firebombings. Their leaders have told the few reporters who have managed to see them that they do not expect the peace talks to succeed. Unless they conclude with a Palestinian flag flying over Jerusalem, the result will not be accepted.

Soldier's Testimony

In mid-1991 a young Israeli soldier revealed something of the tensions and dangers of patrol duty in Gaza in a widely-published letter.[4] He first explained that a briefing to soldiers entering Gaza for the first time finished with the warning that everyone in the Gaza Strip hates Israel and Israelis. Soldiers should be under no illusions about this. They were also warned that they could fire live ammunition only when their lives were in danger. For 'normal' situations they must use tear gas, rubber bullets, percussion grenades, sand pellets and Israeli army-manufactured firecrackers. Specially trained marksmen could fire small plastic bullets at demonstrators' legs. The letter went on:

> We are ordered not to shoot into schools, hospitals or mosques, although hundreds of rocks are thrown at us from these places. The equipment we have is ineffective and fails to deter the stonethrowers. Israeli infantry are trained to attack the enemy but in Gaza we are stoned all day and told not to respond.
> On patrol we get an emergency call on a radio from one of our special units. 'We've grabbed a masked youth but our car has broken down', the

radio crackles. 'Hundreds of them are on us! Get here quick and make a lot of noise on the way! Move!'

Within minutes the whole town knows something is afoot. When crowds in one neighbourhood whistle, this is a signal for other areas to come to the scene or to block army reinforcements. By the time we are on our way, tearing down the narrow alleyways, the whole town is whistling and thousands are on the streets. They are throwing everything at us — rocks, metal bars and parts of car engines.

An even more frantic message is screamed over the radio. 'For God's sake get here! They're surrounding us!'

They are holding off the crowd, using everything — even live rounds fired over their heads. If we don't get there the soldiers will be forced to shoot into the crowd. I grab the soldier next to me who's sitting next to the door. I'm scared that one of the rocks pelting our vehicle will knock him unconscious and he will fall out. I'm praying the driver doesn't crash or stall.

We arrive at the scene within five minutes — the longest five minutes of my life. The soldiers, still holding onto their masked captive, leap into our vehicles and we get the hell out of there. By now we have fired everything except our live bullets. Somehow we get back to our base. Seven soldiers were injured with cut arms and gashes. Mission accomplished. There is a major celebration. All for one local activist!

Great Hopes, Great Despair

On 13 September 1993 Yitzhak Rabin and Yasser Arafat shook hands on the lawn of the White House in Washington in a ceremony that gripped the world's imagination and inspired its hopes. The product of delicate negotiations by a team of Norwegians led by the late Johan J. Holst, the meeting between the representatives of arch-enemies was sponsored by the Clinton administration.

The agreement reached by the Israeli and Palestinian negotiators, known officially as the 'Declaration of Principles', called for a large degree of Palestinian independence in Gaza and the West Bank. Israel would permit Arafat to employ between 8,000 and 10,000 policemen in the territories.

While all peace-loving people welcomed the plan for real peace many others condemned it and openly declared that they would work to wreck it. *Hamas* denounced the agreement, as did certain elements of the PLO, as well as Jewish zealots living in the Israeli West Bank settlements. From outside the country, Iran also declared its intention to wreck the peace, as did *Hezbollah* in Lebanon and Islamic fundamentalist groups throughout the Muslim world.

The greatest blow to peace between Israelis and Palestinians came on 26 February 1994 when Baruch Goldstein, an Israeli of American origin, murdered 43 Palestinians while they were worshipping in the mosque at Hebron. A fanatic, Goldstein was a follower of the bigot Rabbi Meir Kahane, who was assassinated in 1990. He was proclaimed a hero by many settlers but Yitzhak Rabin called him 'a shame on Zionism and an embarrassment to Judaism'.

Even before the Hebron massacre, the former head of Israeli intelligence, Shlomo Gazit, wrote in the *Jerusalem Post*:

'We are now taking the first steps towards an abyss from which there is no return. Israel's problem is not the breach of a consensus which never existed; it is whether the two parts of the people, split in their convictions, can live together and resolve their differences without careering into civil war.'

Gazit was referring to a split between those Israelis in favour of, and those opposed to, aggressive Jewish colonisation of the occupied territories. But following the signing of the Washington peace accords not one but two civil wars began. The first is among Israelis, the one that Gazit feared. The second is among Palestinians. Divisions among them are deep and widening. After the Hebron killings Palestinian demonstrators were chanting, 'Death to Arafat!'. Many fear that his police in Gaza and Jericho will be more repressive than the Israelis ever were.

Many activist members of *Fatah*, Arafat's own organisation within the PLO, have resigned because they say he awarded the best jobs in the new administration to sycophants from Tunisia, the PLO's base, rather than to Gazans and West Bankers who have been at the sharp end of the *intifada* and in conflicts before that.

Hamas guerrillas were saying, early in 1994, that they did not want an open confrontation with their main rival, *Fatah*, but they would resist any attempt by Arafat to prevent them attacking Israelis in the territory. They knew that since September 1993 *Fatah* had been co-operating with Israel's occupation authorities.

The two civil wars are actually in progress. The Israeli government has disarmed only a handful of the 130,000 settlers, who have a strength out of all proportion to their numbers. An Israeli army reservist, writing in the weekly *Ha'olam Ha'ze*, said of his service in Hebron: 'When we had to intervene in a skirmish between the Arabs and the settlers I felt more secure when my back was turned to the Arabs than to the settlers.'

Meanwhile, Palestinians loyal to *Fatah* and Palestinians loyal to *Hamas* are killing one another. In April, Islamic fanatics twice exploded car bombs against Israeli targets, killing 20 people and wounding many more. This tactic was much used in the early 1980s in Lebanon.

The handshake on the White House lawn was a declaration of desperation. It was never likely to produce a peace that would last much longer than it took the ink to dry on the document.

References

1. This is the consensus among experienced Middle East journalists and diplomats. 'Every facet of Palestinian politics would change radically if Arafat were to disappear', a Turkish diplomat said.
2. The Israeli newspapers were full of angry letters from members of the public and in the Knesset there were furious exchanges among the members and ministers.
3. Reported by Michael Sheridan, *The Independent's* correspondent in Israel, 21 June 1991.

4. Some people might see this letter as a form of propaganda to enable the military to stress the fire discipline imposed upon its soldiers and the dangers which they face from murderous mobs. Even if this is the case, I can confirm that the letter accurately describes the type of patrol referred to.

War Annual No. 5 describes Defence Minister Arens' tactics to contain and defeat the *intifada*; the Temple Mount riot and shootings; PLO diplomatic and military failures.

Kurdish War of Independence

NEW PHASE
Background Summary

The Kurds' right to create their own state was granted in the Treaty of Sèvres, following the First World War, between the Allied victors and the defeated Turks. Unfortunately, the Sèvres Treaty was never ratified, being superseded by the Treaty of Lausanne. This time the Kurds were not recognised and the area that they claimed was carved up, with the largest portion going to Turkey. After the Second World War, Iranian Kurds, with Soviet support, set up a 'Republic of Mahabad' but it lasted only one year.

The Kurds face two major obstacles to statehood. The first is that they occupy a region which straddles the borders of Iraq, Iran, Turkey, Syria and the Soviet Union, all of which oppose the idea of a separate Kurdish state. The second obstacle is that the Kurds are without national unity; they have always owed primary allegiance to their tribe.

Much affected by the Iran–Iraq war, the Kurds hoped that the ceasefire in 1988 would bring them some relief from being considered everybody's enemy. But Saddam Hussein turned against his own Kurds in March 1988, using chemical weapons against them. More than 200,000 were killed or injured. Saddam's intention was to crush forever the traditional demand for self-determination and the Kurds' spirit to resist. However, Kurdish resistance continued in Iraq, Turkey and Iran, especially in Turkey.

The 'Gulf Crisis' of August 1990, when Saddam Hussein invaded Kuwait, greatly affected the Kurds. Their oppression by the Turks had been largely ignored abroad and when Turkey joined in the UN sanctions against Iraq, its military activities against the Kurds were even less likely to excite attention in the West. The Kurdish leaders believed that they would have a chance of gaining statehood if Saddam were to be killed during the UN–Iraq War. When this did not happen and Saddam remained securely in control, the next phase of Kurdish suffering began.

The Wars Within Iraq

When the ceasefire for the UN–Iraq war came into effect on 28 February 1991 President Bush explained, yet again, that the purpose of the war had not been to defeat Saddam Hussein or occupy Iraq. The coalition forces would leave southern Iraq immediately after liberated Kuwait was considered secure. He suggested that the future of Iraq was in the hands of the Iraqis and he strongly implied that those who saw Saddam as a threat to the future of their country might rise against him. The Shia Muslims of southern Iraq and the

Kurds of the north, all Iraqi citizens, accepted this 'invitation'. Some of their leaders and many of the ordinary people believed that their uprisings would be helped by the Americans and British, if not by the entire coalition forces. This did not happen and two civil wars raged as Saddam's forces, which were still organised and powerful, used ruthless tactics to put down the rebellions.

The Shia War in the South

The Shias of Iraq may form a majority of the people of Iraq but they have never been as organised as the Sunnis, who control the army, the government and much of the bureaucracy. The Sunnis, who are well aware of the spiritual affinity which the Iraqi Shias have with the Iranian Shias, have always seen to it that the Shias remained politically weak. Despite the second-class citizenship imposed upon them, the Shia community gave more than a million of its young men to the national army for the eight-year war against Iran.

In March 1991 the Shias began an *intifada* against Saddam. Their nominal leader was Grand Ayatollah Abul Qassem al-Khoei, an ailing 95-year-old, but their real leaders were younger men, many of them clerics from the holy cities of Najaf, Kerbala and Qum.

Under the leadership of these clerics and officer deserters from the Iraqi army, the uprising became intense. Saddam's units were ambushed and attacked, officers were murdered, tanks were stolen and there was widespread sabotage. The Shias expected help from their co-religionists across the border, at least in the form of food and supplies, but it did not reach the scale they expected and needed.

In March, Saddam's security police seized Ayatollah al-Khoei and he was paraded on Baghdad television, as the Western hostages had been. The British Foreign Office expressed its 'outrage' over the Ayatollah's treatment in a letter sent to his family in London. 'The government is conscious of the plight of the Shias in Iraq. We have not forgotten them, despite all the attention currently being focused on northern Iraq.' This was a reference to the Kurds.[1]

In fact, neither the British nor any other Western government was concerned about the Shias and their *intifada*. Everybody was conscious of the fundamentalism of the Shias and there was a great fear in the West that, should they succeed in removing Saddam and his supporters, they would set up an extreme Islamic revolutionary government in Baghdad similar to that in Tehran. For some Western statesmen even Saddam Hussein was preferable to that. Also, the vast majority of Shias were out of sight. They were trapped in the cities and towns of the south where Iraqi troops were operating under the slogan 'No Shias after today'. No international supervision let alone control, was possible.

Saddam insulted the Shias and their faith. A commentary in a government daily newspaper stated: 'The leader the Shias deal with is called a *sayyed*. These *sayyeds* take the money of the poor farmers, make them kiss their feet and bend to the ground for them.'

Saddam was intent on revenge. Any Shias trying to avoid his retribution had to cross into Iran, which became increasingly difficult because of army patrols, or take the Jordan road that runs through Baghdad. This was even more dangerous. The Iraqis bulldozed the cemetery in Najaf because it had been a refuge and rallying point during the *intifada*. They sealed off the shrines of Najaf and Kerbaba from worshippers and systematically destroyed the surrounding areas. Three schools of Najaf University, a 1,000-year-old centre of learning, were damaged and a fourth bearing the al-Khoei name was razed.

With international political support conspicuously lacking, Youssef al-Khoei, grandson of the Grand Ayatollah, visited the Vatican on 8 May in an attempt to win papal support for the Shias. He also met Prince Sadruddin Aga Khan, the UN co-ordinator for Iraq refugees, and reported to him that 300 religious teachers and 500 of their pupils had been rounded up and disappeared. In fact, they had been massacred.[2]

Many thousands of desperate Shias fled to the American lines in the buffer zone between Kuwait and Iraq. The numbers strained the US military's ability to help them and enforce the ceasefire at the same time. Some refugees were soldiers-turned-rebels whose units had been crushed and now had nowhere else to turn. Some were deserters from the Iraqi army and Republican Guard who could no longer carry out orders for attacks against their own people.

The refugees choked US checkpoints set up along the Nasiriyah highway which runs parallel to the demarcation line agreed at the end of the war. For a time the Americans provided no facilities for refugees and turned back soldiers who wished to surrender. After pressure from the International Red Cross, the Americans set up refugee camps, mobile hospitals and POW-holding areas in the stretches of desert that straddled the highway. They also opened a refugee camp and food distribution centre in the Iraqi border town of Safwan. These facilities attracted even more refugees. Some checkpoints resumed acceptance of the surrender of Iraqi soldiers on the insistence of US counter-intelligence officers.

The US policy was not to help the rebels, a strange attitude in the light of President Bush's urging them to fight against Saddam Hussein's regime. However, in some places, such as the checkpoints around Suq al Shuyukh, Iraqis moving north of US positions were allowed to keep their weapons; if they moved south across the US line of control they had to surrender them. The type of weapons taken here indicated the ill-preparedness of the Shias for military operations against the Iraqi army. There were old AK-47s, bolt-action rifles manufactured in the 1940s and hand-grenades.

Wherever the rebels were under former army officers or senior NCOs they fought well.[3] The instructors gave them rudimentary training in camouflage, ambush positions and fire and movement, and as a result small-scale defeats were inflicted on the army. However, each such rebel victory was followed by army atrocities against Shia civilians.

While refugees in their hundreds of thousands fled into the mountains bordering Iraq and Iran, the fighting men retreated to the Haw al-Hammer marshes, where many civilians were already hiding and living in great hardship. From here the hardier fighters made forays against Iraqi military targets but they were the acts of a dying uprising. Iraqi army units encircled the marshlands in Operation al-Anfal. They were briefly withdrawn in July when Prince Sadruddin Aga Khan visited the marshes and asked the Iraqi government to reduce the number of army units so that food and water could reach the refugees. Following the Prince's departure, the army resumed its positions, preventing people and supplies from moving in or out. UN officials discounted reports, which had emanated from Baghdad, that up to half-a-million refugees were in the marshes and put the figure as 'closer to 30,000'.

The narrowest gap between the marshes and the Iranian border is 30 miles. By July most Shia hit-and-run raids were being mounted from Iran and the Iraqi army built numerous watchtowers to watch for rebel movement.

During the Iran–Iraq war the marshes, which are formed by the Tigris and Euphrates north and west of Basra, were a refuge for deserters and dissidents. The army drained part of the marshes in order to contain and then capture these men. This process was still going on in 1992 and as the water level goes down the villagers of the marshes cannot fish. They are then forced to move into the cities of Basra, Nasairiya and Amara. Troops can then more easily control the vast swamps as well as the town Shias.

The Shias' *intifada* was a desperate and a brave attempt to end the tyranny which Saddam Hussein had imposed upon them. It failed largely because the Western allies and equally the Tehran government did not support it. Certainly the leading Shias of Iraq would like to see Iraq become a fundamentalist Islamic republic, a fact which influenced the American decision to stay out of the fighting. However, so many counterbalances existed in Iraq that it is unlikely that they could have gained control in Baghdad. Had the West simultaneously backed the Shias, the Kurds and the dissident anti-Saddam Sunnis, a government of national unity might have supplanted Saddam's regime.

Casualties in the Shia *intifada* are unknown but UN officials who visited southern Iraq estimate Shia losses to be 40,000 dead and 'many' wounded. It is impossible to separate the civilian dead from those who died while fighting against the Iraqi army. Baghdad has given no figures for army casualties and Shia claims relating to them are so wildly inflated that they have no credibility.

According to a government official who could no longer obey the cruel orders of his Ba'ath Party superiors and escaped to Jordan, children had been tossed from helicopters as a warning to their parents not to become involved in activities against the government. The head nurse in a hospital at Kerbala had her breasts cut off in public after treating wounded Shia fighters. When a local police commander apologised for having fired in the air to disperse a demonstration, Saddam Hussein himself responded on national television: 'In the air? Are you celebrating a wedding? Shoot them in the head.'[4]

The Kurdish Uprising

While the United States was holding victory parades, Saddam Hussein was getting ready to crush the Kurdish *peshmergas* or guerrillas who during the Allied offensive had taken the opportunity to attack Iraqi garrisons. The Kurds were not engaged in open warfare at this time but their leaders believed that they should show Saddam's international enemies that they were helping in the campaign against him. In any case, they believed that the land offensive which had followed the air onslaught would depose Saddam Hussein one way or another; they were astounded when the offensive stopped, leaving him still in

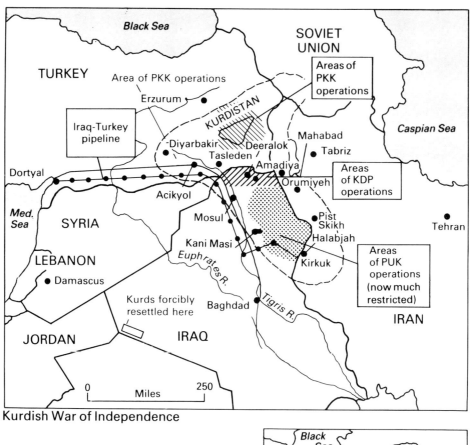

Kurdish War of Independence

power. However, their spirits lifted again when President Bush encouraged them to rise against Saddam. They assumed that the Allied forces would at least protect them against Iraqi air attack.

On 27 March, Saddam swore in his new cabinet and told the ministers that they had four to six months to 'prove' themselves. The new Interior Minister, Ali Hassan al-Majid, and the Defence Minister, Saadi Toma Abbas, were not present at the Baghdad ceremony. They were already in the north to organise the defence against the Kurdish rebels. Majid, who as commander of northern Iraq had orchestrated the chemical attack against the Kurds at Halabja, was in fact in Mosul.

The Kurdish fighters were doing well. They had captured an air base at Khalid near the oil city of Kirkuk and destroyed three fighter-bomber aircraft, probably Su-22s, and seven M-8 helicopters. Bases in themselves were of no use to the Kurds, who had no pilots, but they hoped that international agencies would use them to distribute emergency rations. Massoud Barzani who commanded the Kurds on this front, wrote to the UN Secretary-General asking for the UN and the International Red Cross to distribute food and other aid to all Iraqis, not just to the Kurds.

As listed in *War Annual 5* the main Kurdish Political Groupings are:
- Kurdish Democratic Front (KDP) led by the Barzani family, and normally resident in Iraq.
- Patriotic Union of Kurdistan (PUK) led by Jalal Talabani, largely resident in Iran.
- Kurdish Workers Party (PKK) of 'Apo' Abdullah Ocalan, which was outlawed in Turkey and found refuge in Syria.
- People's Liberation Army of Kurdistan (ARGK), which operates mostly in Turkey.
- Less important groups are the Socialist Party of Kurdistan (Pasok), the Iraqi Dawe Party (a Shia Muslim group) and the Turkish Workers Party of Kurdistan. There is also a breakaway group of the KDP founded by Abdorrahmen Qassemlou. In 1988 the PUK and KDP formed a tactical alliance, the National Front of Kurdistan (NDF).

Jalal Talabani, of the Patriotic Union of Kurdistan (PUK) who had returned from exile on 26 March, said that he hoped to use Kurdish officers who had defected from the Iraqi army to train the thousands of Kurdish fighters who had seized weapons from government installations but who were new to combat.

The Kurds' prospects were good at this time, although up to 1,000 people had been killed the previous week during Iraqi air bombardment of Kirkuk and Dohuk and as many again in attacks on Kifi, Kalar and Tuz Khurmatu to the south. The Iraqi News Agency reported on 27 March that Saddam's troops were 'mopping up pockets of saboteurs in the northern areas'. In Washington, the Pentagon stated that the Iraqi forces were still too busy fighting Shia insurgents in the south to pay much attention to the north. This was wrong. The Iraqi forces were quite strong enough to deal with two insurgencies at the same time.

Saddam Hussein judged that the Western powers would not intervene to help the Kurds and saw his opportunity to expel large numbers of them by force. In doing this he was maintaining a policy established in 1980 to rid Iraqi Kurdistan of the troublesome Kurds and replace them with more malleable Egyptian and other Arab immigrants. His armoured units and combat engineers, supported by infantry, moved against many Kurdish towns and villages, a number of which had suffered severe damage during earlier pogroms. Now they were flattened. The Kurdish civilians fled *en masse* to the mountains, despite appalling conditions caused by the bitter winter weather. During the first two weeks of April the *peshmergas* were weakened as an offensive force as many units were needed to escort the fleeing civilians.

Apart from vengeance, Saddam had two main objectives in turning on the Kurds so violently. One was to tighten his control of 'Kurdistan' by pushing rebellious populations right out of Iraq. The other was to use refugees in effect as an offensive weapon, by forcing them across frontiers in numbers large enough to disrupt the societies of neighbouring countries.

The US government warned Baghdad that government forces must not venture northwards across the 36th Parallel under threat of having their aircraft shot down and their ground forces attacked. Nevertheless, the Iraqi attacks on refugees continued. The major Kurdish city of Sulaymaniyah, which the Kurds hoped to retake from the Iraqis, was south of the 36th Parallel so the American threat of counter-attack did nothing to aid the Kurds here. Talabani's men set up camps in the hills around the city but they were terribly vulnerable to attack by helicopter gunships.

While the Iraqi armed forces were being systematically brutal towards the Kurdish refugees, Iraqi officials told the UN envoy in Baghdad that the government would co-operate with UN efforts to alleviate the plight of the refugees. This was characteristic Saddam-inspired hypocrisy. Saddam offered an amnesty to those Kurds who would accept his offer of 'peace and harmony' and the Israeli media reported that 'thousands of delighted Kurds are pouring back to the northern towns'. The Iraqi news agency said: 'The returning families greeted the singular and wise leadership of President Saddam Hussein in his heroic and decisive confrontation of the evil plot to strike at the unity of the Iraqi people and encroach on their great achievements.' The 'plot' was said to be the work of the United States, Britain and their 'lackeys'.

Even as the Iraqi officials were talking of better times for them, hundreds of thousands of Kurds were struggling over the freezing mountains for dubious refuge in Turkey or Iran. All over the world television audiences saw the refugees suffering from cold, hunger, exhaustion and privation. Dr. Gerard Salerio of the voluntary organisation *Médecins du Monde* (Doctors of the World) provided eyewitness testimony of the Kurds' agony.

> There are sometimes up to 40 people living under the same tent but these are not even tents — they are stretched blankets. People are too ashamed to relieve themselves during the day so they do it at night-time, between

the tents. There is no hygiene anywhere. Every day 20 children are buried between the tents. Older people are dying too, so are younger adults. They are dying, dying even as I speak and one doctor serves 100,000 people.[5]

Operation Provide Comfort, the US military's effort to bring relief, began on 15 April with the objective of supplying at least one meal a day to 700,000 Kurds for a month or so, until the UN and private relief organisations could take over. Distribution of such goods as reached close to the Turkish frontier was held up by incompetence and indifference. Twenty-one planeloads of relief supplies were delivered to the eastern Turkish town of Diyarbakir but much of the material did not get beyond the airport. Other supplies rotted in the rain aboard trucks stuck on the muddy roads of south-eastern Turkey.

The refugees were dying at the rate of 1,000 a day but the Turkish government refused to allow any Kurds to cross the border; they feared that they would join the Turkish Kurds in forming a political bloc demanding more autonomy than Ankara was prepared to grant. British Prime Minister John Major developed an idea first suggested by the Turkish President Turgut Ozal for a stopgap solution. This was the creation of UN-sanctioned 'enclaves' — later called 'safe havens' — where the refugees would be protected from attack by Saddam's forces.

Some members of the UN Security Council — the Soviet Union, India and China — feared setting a precedent of intervention in what have always been considered internal affairs. They saw the precedent being applied to minorities in their own lands. The US administration saw little chance of getting the resolution through the Security Council and, backed by Britain and France, the Americans unilaterally created safe havens, in the 10 per cent of Iraq that lies north of the 36th Parallel.

Allied Troops in Northern Iraq

Over the week-end of 27–28 April 1991 American, British and French troops moved into northern Iraq, an area the Allies had largely left alone during the war, to begin *Operation Haven*. They quickly built seven tent cities, each to house up to 100,000 Kurds and the refugees were gradually brought down from the barren, freezing and wholly uninhabitable mountain slopes. In the camps they were given adequate food, water, shelter, sanitation and medical care. Most important, the camps were protected by 10,000 soldiers from the US, 5,000 marines and soldiers from the UK, 1,300 from France and 1,000 each from Italy and the Netherlands.

Commander of the relief effort was US Army Lieutenant General John Shalikashvili, while the British commander was Major General Robin Ross of the Royal Marines. Shalikashvili at once called Iraqi officers to a meeting near the border town of Zakhu, to warn them to keep their 30,000 troops in the region away from the camps. The US and British governments hoped to turn over protection of the refugee settlements to a UN peace-keeping force. Many observers assumed that Allied forces would have to stay until Saddam Hussein departed, but since Washington had no strategy for forcing him out, that could mean maintaining garrisons for years.

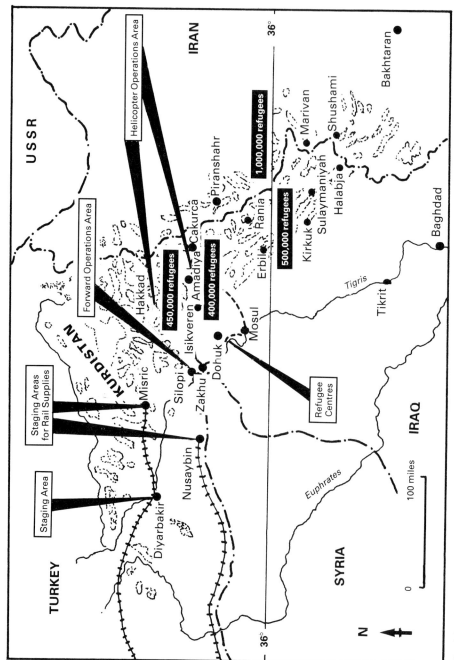

Saddam's War against the Kurds: Allied Protection

The marines of 45 Royal Marine Commando led the Allied push towards Amadiyah, part of the plan to double the size of the Western safe haven to 80 miles east of their first camp at Zakhu. At a meeting with Iraqi representatives, a US Cobra helicopter gunship hovered 20 feet over the Iraqi special forces camp. Two light assault vehicles of the US Army's Second Light Armoured Infantry Division drew up at the camp, their 25mm cannon aimed at the troops inside the camp fence.

The Royal Marines neared the camp on foot, skirting the perimeter of Saddam's summer palace, several miles away. It had been decided that the palace's resident Presidential Guard could remain, but only with light weapons. The Republican Guard special forces were ordered to close down their camp and get out.

While the troops of *Operation Haven* were still hard at work on behalf of the suffering Kurds an extraordinary event took place. Saddam Hussein and Jalal Talabani had a formal meeting at which they smiled at each other, embraced and kissed. The coming together of the two leaders with such apparent lack of animosity was seen on television around the world and startled even veteran Middle East observers. After all, Saddam had once said that he would run a sword through the rebellious Talabani rather than permit him to return to Iraq. Talabani was the leader of people who had been betrayed, gassed, bombed, shot and forced into exile by Saddam Hussein.

After five days of talks, the two sides tentatively agreed that in exchange for the Kurds ending their uprising, Baghdad would give them some form of autonomy in northern Iraq. Even if an armistice held for a time, no experienced analyst expected it to bring lasting peace for the Kurds. The allies did not pause in their efforts to establish a safe haven. If a final pact prompted the Kurds to return home, it would relieve the Allies of their enormous difficulties in trying to aid the refugees without further friction with Baghdad.

The simple truth of the Talabani–Saddam embrace was that both leaders were trying to buy time. Talabani's willingness to meet Saddam could be interpreted as an acknowledgement of Kurdish military failure, and certainly in terms of Middle East politics Saddam gained the most from the much-publicised meeting.

Why the Kurdish Revolt Failed

The scale of the Kurdish uprising surprised even its leaders and they were overwhelmed by the speed and magnitude of events. The initial successes were gratifying, for practically the whole of the rural areas fell into their hands as well as the northern Kurdish cities, Kirkuk, Erbil, Dohuk and Zakhu. The 20,000 lightly-armed guerrillas who formed the backbone of the uprising found that they had taken on the administration of more than three million people. It was an impossible task.

Many towns and cities fell with little fighting; only at Kirkuk was the action really fiece. Nevertheless the Iraqi Army's First and Fifth Corps, the garrison of the north, laid down their arms. The rebels were joined by thousands of Kurdish soldiers deserting from the Iraqi Army and from the Kurdish militia, the government's 'national defence brigades'. When the 200,000 Kurdish

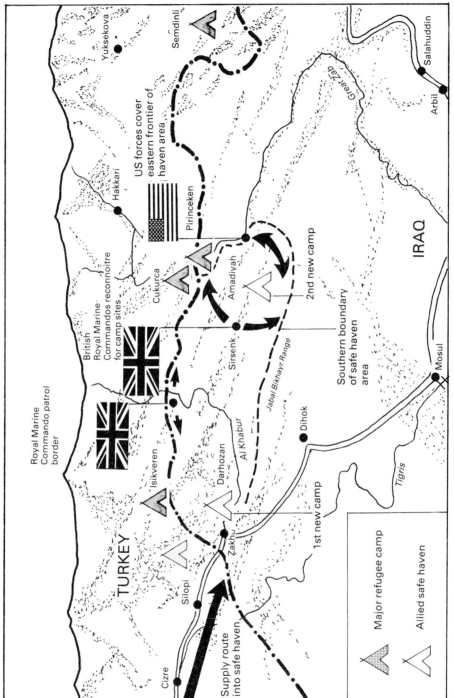

'Happy Valley' Safe Haven

militiamen defected, it appeared that practically the whole of Kurdish society had turned against Saddam.

The hard-core rebels, the 'originals', now had an immense number of men and weapons to organise. Simultaneously they were trying to administer the cities they had taken over — but without adequate communications, food or industrial fuel. The tasks were overwhelming, but the Kurdish leaders were jubilant over their successes. At this point Saddam, using the Republican Guards as his main strike force, hit back with a sustained violence that shocked the Kurds. They had only their Kalashnikovs to fight against artillery, tanks and helicopter gunships.

Western intelligence officers had told the Kurdish leaders that the Republican Guard divisions had been so weakened by the Allied bombing and by the ground offensive that they no longer posed a serious threat. The Kurds believed this outrageously incorrect intelligence. They found another difficulty — the vast floods of refugees who got in the way of the *peshmerga* as they prepared to do battle with the Iraqi forces. Many *peshmerga* were caught up in the stampede or left their posts to help their families. According to Jalal Talabani, he was left with a mere 15 men to defend the front in the mountains north of Sulaymaniya. Many of those who left their units later returned but by then all the major cities which had fallen to the rebels were back in Saddam's hands.

In essence, and ironically, the uprising failed because it had succeeded too well. The *peshmerga* were guerrillas, accustomed to fighting on their own terms in rough terrain. They could not be expected quickly to adapt to street-fighting and to manoeuvre battles in the open.[6]

The Kurdish Recovery

On 18 July, the Kurds under Sheikh Salah Hafid captured Sulaymaniyah, a city of 350,000 people. A day of fighting left 250 dead, 600 wounded and 2,500 government troops prisoner. The *peshmerga* won control of almost all the city but it remained ringed by forces of Iraqi infantry, armour and artillery.

Within days of the victory, 40,000 Kurds fled from the city, convinced that there would be a second round of fighting. The Allies were also alarmed because an offensive would force them to decide whether or not to use their rapid-reaction force based in Turkey to stop the Iraqi army from gaining control.

The city remained nominally under the joint control of the government and the *peshmergas*. An agreement reached by the Iraqi governor, Muhammad Najmuddin al-Naquishbandi, and Sheikh Hafid specified that 500 *peshmerga* and 500 soldiers and police would patrol the streets together.

Kurdish disunity showed itself again throughout the summer. Massoud Barzani's party was convinced that there had never been a better chance for the Kurds to extract an autonomy deal from a weakened central government and that any successor to Saddam would have to abide by it. But Jalal Talabani's party, despite his earlier embrace with Saddam, believed that a hasty agreement could lead the West to abandon the Kurds altogether. For Talabani, a better course was to let negotiations simmer until 'natural causes' blew Saddam away and brought to power somebody in whom the Kurds could have a little more trust.

Barzani and his party spent 42 days in Baghdad and returned to Kurdistan with a draft agreement. They were satisfied that they had got everything that Baghdad was prepared to give. The eight political parties in the Iraqi Kurdish Front prepared to debate the draft, a process which most observers believed would take many months. In fact, negotiations collapsed on 6 October 1991. Saddam's troops begun to bombard Sulaymaniyah and other Kurdish towns, Kifri, Kalar and Maydan. Many hundreds of civilians were killed in the shelling and tens of thousands once again fled for the hills, where the *peshmerga* began to mass.

Kurdish leaders said that Baghdad had ordered the shelling to coincide with the visit to Washington of Jalal Talabani, and in the belief that, with the world distracted by Iraqi weapons secrets, the Kurds could be attacked with impunity. This seemed to be the case. The US and its allies expressed concern but refrained from intervening.

The Turkish Front

The Kurdish uprising in Iraq and the consequent flight of refugees, followed by the drama in the mountains, almost completely overshadowed the activities of the Kurdish Workers Party (PKK) and its efforts to set up an independent Kurdistan in south-eastern Turkey. The outlawed PKK's guerrilla activity against the Turkish security forces had never stopped. Largely for this reason, Turkish involvement in the international aid being directed towards the suffering Kurds of Iraq was minimal. The Turks agreed to allow some of their bases to be used for the distribution of food and shelter but they did nothing else to support the international relief effort.

The PKK's activities against the Turks reached major proportions early in August 1991 when guerrillas crossed the border to attack an an army post. Nine soldiers were killed in this raid and another seven were taken prisoner. In retaliation, Turkish troops and F-4 and F-104 planes attacked Turkish Kurdish bases inside northern Iraq. This was the fourth time since 1983 that Turkish forces had crossed the border in hot pursuit of raiders. They did this under a Turkey–Iraq pact of 1984 that allows the armed forces to pursue Kurdish rebels into each other's territory for a distance of six miles.

A statement from the Turkish General Staff, quoted on Turkish radio, said that fighter-bombers had flown 132 sorties against Kurdish targets in Iraq. In addition, troops flown into the area by helicopter had caused 'heavy damage' to the Kurdish bases. The communiqué, which was signed by the armed forces chief, General Dogan Gures, concluded: 'It should never be forgotten that the Turkish army is ready to punish severely and decisively those responsible for all kinds of actions and treachery against our country and the security of our nation.'

Soon after this operation Turkey announced that it would occupy a three-mile deep buffer zone in northern Iraq to prevent guerrillas from infiltrating into its territory. This was the most dramatic move that Turkey had taken in seven years of fighting Kurdish separatists. It could put Turkey into conflict with Iraq and raise tensions with the US-led allied rapid reaction force of 5,000 stationed at its border with Iraq to defend Iraq's Kurds. The force includes a Turkish contingent.

The Iraqi Kurds deny any formal alliance with the Turkish Kurds, whose campaign had cost 3,200 lives to the end of 1991.

Open War in Turkey

The Kurdish new year, known as *Nowruz*, was the setting for open warfare between the PKK and the Turkish army in the latter half of March 1992. The PKK had proclaimed *Nowruz* as the beginning of an all-out war against Turkey and publicly demanded that all Kurds should be prepared to fight 'to the last man.'

The fighting over the weekend of 20–22 March was the most violent in the PKK's eight-year struggle. For the first time, the army deployed main battle tanks in Cizre, in the south-east, which is home for about half of Turkey's 12 million Kurds. Guerrillas armed with assault rifles opened fire on police and troops. The PKK's amnesty for collaborators ended on 21 March when its killers hanged three 'traitor' Kurds, their mouths stuffed with money, from lamp-posts in Cizre.

The fighting showed the real depth of the conflict in the south-east. The Turkish government has largely concealed the problems from the people but on this occasion they saw much of the fighting on television and they heard about police brutality against the Kurds. Police death squads have killed many Kurdish nationalists. For the first time, warfare has spread outside the area known as Botan, a large triangle with sides about 200 miles long, along the Iraqi–Syrian borders.

Diplomats in Ankara say that the security forces in the area were trying, over the *Nowruz* period, to draw the PKK into showing their strength prematurely. In fact, the group has about 10,000 fighters. It is clear that in the Botan area the PKK has great support and that the army will be using tough methods to deal with the Kurdish irregulars. The Prime Minister, Seleiman Demirel, a liberal-minded man, frequently appealed for calm and for 'sensible negotiations' but his pleas fell on deaf ears on both sides of the conflict.

Faint Hope from New Prime Minister

Many Kurds believed that the election of Turkey's first woman prime minister, Tansu Ciller, in 1993, would bring better conditions for the Kurds. But Ciller's first foreign policy success was to induce the German government to ban the 30 Kurdish organisations in Germany. Turkey then began to press its other European allies to take similar action.

Western support for Turkey comes at a time when Turkey's strategic role on NATO's southern flank has become important, following concern over Russia's dominant role in the independent states of the former Soviet Union. Western diplomats insist that the Ciller government will have to show a more conciliatory attitude towards the Kurds. However, the Turks know that they can, literally, get away with murder.

During 1993 and 1994 Turkish security forces stepped up their cross-border and attacks against rebel bases in northern Iraq. Simultaneously, Iran, always seeking an ally, has courted Turkey by inviting senior security officers into Iran to search for rebel camps. Then troops from the two countries will unite to destroy 'these vermin'. Syria also announced an official ban on the PKK, which has a base in northern Syria. This may not mean a great deal as Damascus has 'banned' the PKK on other occasions.

For the Kurds, 1994 has been as desperate as 1993. Progress towards greater autonomy was stalled by the West's unwillingness to upset Turkey and by the desire of Arab and other states at the UN not to dismember Iraq. Iraq's embargo of goods of all kinds getting into Kurdistan has holes in it but it still causes great hardship to the Kurdish people.

References

1. A Foreign Office official told me privately: 'We must never forget the terrible crimes committed by the fundamentalist regime of Ayatollah Khomeini when it came to power in Tehran. The Shias of Iraq revere the memory of Khomeini and almost certainly if a fundamentalist regime took over in Baghdad it would be as ruthless as that in Tehran was. Also, there would then be a Tehran–Baghdad axis of fundamentalism. The dangers from that do not bear thinking about.'
2. The remains of numbers of them were found months later. They had been hacked to death.
3. This assessment was made by American intelligence officers who had the opportunity to interrogate former Shia fighters and Iraqi soldiers who had fought against them.
4. Reported by Julie Flint in *The Observer* 9 June 1991.
5. Dr. Salerio was speaking to George Church of *Time Magazine*; 22 April 1991
6. I have discussed this analysis with several Kurdish political leaders, most of whom agree with it. However, several said 'the main reason we failed was that the West allowed Saddam to attack us'. They showed great bitterness when making this comment.

Looking For Peace in Lebanon

THE SYRIAN COUP

Background Summary

Lebanon has not known total peace for many decades but before 1975 conflict was intermittent and fighting was tribal. The various ethnic and religious groups had managed to live in a kind of peace based on the conviction that a prosperous Lebanon had something to offer all its inhabitants. In 1975 a 'civil war' began.

More accurately, it was a war between the Palestine Liberation Organisation (PLO) and the 'Lebanese Forces', the name adopted by the Maronite Christian militia. The PLO had settled in southern Lebanon after being bloodily expelled from Jordan in 1970 and it became a state within a state in a region known as Fatahland, named after Yasser Arafat's PLO faction. When the Christians were losing the war, the Syrian army entered Lebanon to aid them. Later, the Syrians changed sides back to the PLO. Syria had traditionally claimed that Lebanon was actually part of Syria and its leaders intended to recover it.

Shia Muslims of the south, Sunni Muslims of the centre-west Lebanon and the Druse of the Shouf Mountains entered the conflict, which became ever more complex. An estimated 100,000 Lebanese were killed, most of them in massacres, in the period 1975–83.

Retaliating against PLO raids in Israel, Israeli troops entered Lebanon in *Operation Litani* in 1978. This drew into Lebanon the United Nations Interim Force in Lebanon (UNIFIL). When raids, cross-border rocketing and shelling continued, Israel launched *Operation Peace for Galilee* (6 June 1982) and defeated both the PLO and Syrian forces.

Ferocious fighting took place between Maronites, Druse, Shia Muslims, Sunni Muslims and Syrians, in shifting alliances. In 1987 two wars took place simultaneously. In one, the *Hezbollah* (Party of God) extremist Islamic movement fought the Shia Militia *Amal*. Then, these two organisations also fought the Israeli forces and the Israeli-financed South Lebanese Army (SLA) in the southern security zone declared by Israel.

In March 1989 General Michel Aoun began a Christian 'War of Liberation' against the occupying Syrian forces and their Lebanese Muslim allies, such as *Amal* and the Druse. During the year thousands of people were killed in artillery duels, while notables, such as President Rene Moawad and the Mufti of Lebanon, were assassinated by car bombs. Late in 1989, an Israeli column raided deep into Lebanon to destroy the headquarters of the Lebanese Communist Party (LCP).

A new war began on 31 January 1990 between General Aoun's army and the Phalangist militia, the Lebanese Forces, under Sami Geagea. By April, 2,500 people had been killed and 7,500 wounded. Strange alliances developed, with Geagea receiving aid from both Iraq and Israel, who were bitter enemies, and Aoun receiving aid from Syria, against which he had fought a 'war of liberation'. The inter-Christian combat was more savage than almost any other phase of the Lebanese fighting and casualties mounted — 3,600 dead by July and 11,000 wounded.

The Phalangist Lebanese Forces formed an alliance with President Elias Hrawi's administration and Aoun's mutinous career came to an end on 13 October 1990 when his opponents crushed his army. Aoun found sanctuary in the French Embassy.

While the two Christian armed forces and their allies were fighting for supremacy in the Maronite enclave of Lebanon, two Muslim militias were engaged in an equally bloody war in southern Lebanon. From the beginning of 1990 Iranian-backed *Hezbollah* was fighting Syrian-supported *Amal* for control of Lebanon's 1.5 million Shia community. Another party to this was the PLO force, which was ordered in by Yasser Arafat to keep the rival Shia Muslim factions apart and to stop *Hezbollah* from gaining control of the area. The Palestinians, who are mostly Sunni Muslims, were not interested in stopping the Shia groups from engaging in reciprocal slaughter. It was merely that Arafat saw the pro-Iranian fundamentalists as a greater danger than *Amal*. In this, Arafat and Israel had the same objective.

Yet another armed conflict erupted. Yasser Arafat's *Fatah* set out to crush the followers of Abu Nidal, the most notorious of all Palestinian terrorists. A three-day battle in the vast Ein el-Hilwe shanty town, near Sidon, resulted in a *Fatah* victory. Civilians long accustomed to war said that the Ein el-Hilwe fighting was the most terrible they had seen.

The War in 1991-92
'The Rape of Lebanon'

President Assad of Syria welcomed the UN war against Iraq as his best opportunity in 20 years to make progress towards Syria's control of Lebanon. Syria and Iraq had been bitter enemies for years, partly from ideological differences, partly because of disputes over territory which both claimed but, perhaps most importantly, because they both sought leadership of the Arab world, after Egypt made peace with Israel and was ostracised by the rest of the Arab states.

By sending a token force to help defend Saudi Arabia and to join the coalition against Saddam Hussein following his invasion of Kuwait, Assad at a stroke made himself popular with the Bush administration and with the Saudi royal family. This meant that while the war against Saddam Hussein was in progress Assad was tacitly given *carte blanche* authority to do what he liked in Lebanon.

In the fire and fury of the Iraq war, the protests of the Lebanese government and people, of Israel and probably all the Arab states went unheard. The US ambassador in Beirut, Ryan Crocker, had authority from Washington to 'intervene intellectually', but he could make no threat of US military action. Crocker tried to induce President Hrawi and all other warlords in Lebanon to drag their feet in the face of Syrian pressure. Crocker was hoping that once the Iraq war finished President Bush and his administration would rescue Lebanon from its predicament.[1]

Crocker's brave endeavours were originally matched by Assad, who sent Deputy President Abd al-Halim Khaddam to West Beirut on a 'courtesy visit' that was, in fact, a stand-over exercise. President Hrawi was invited to Damascus and on 22 May he and Assad signed 'the Damascus Treaty of Fraternity, Co-operation and Co-ordination'. The fiction of this agreement was that Syria recognised Lebanon's independence; the fact was that Lebanon became a puppet regime. When the elderly Maronite Christian leader, Raymonde Edde, returned from exile in Paris he called Lebanon 'a Syrian colony'.[2]

The articles of the Damascus agreement make it feasible for Assad and his successors to incorporate Lebanon into the Syrian army's area of deployment. Assad and his Generals realised, after their failure to conquer Israel in the 1973 war, that they needed a new border, with Israel, in addition to that on the Golan Heights. The Lebanon–Israel frontier is that border.

The Syrian army maintained 25,000 troops in Lebanon until the signing of the Damascus Treaty. Since then it has sent another 13,000 personnel into the country.[3] This army of occupation could be quickly and massively reinforced from the Syrian army of 300,000 and air force of 40,000. In addition, there are reserves of 400,000 who could be rapidly drafted, since Syria is permanently on a war footing.

Once in control of Lebanon, Assad had interesting possibilities to develop the war he has always planned to fight against Israel. Several developments occurred within weeks. *Hezbollah* forces moved down the Baalbek valley towards the southern security zone; PLO elements which had previously retreated from Tufah to Tyre moved back again; many more artillery pieces appeared within range of Israel's security zone. The Syrians asked the Lebanese to build tank traps and other obstacles along the Israeli border.

Tanks and helicopters which the Syrians had seized from the Lebanese Forces (the Christian militia) were handed over to the Lebanese army but that army is subject to Syrian orders.

The Syrians did not get hold of all the Lebanese Forces' heavy armaments. A blue-painted cargo ship, *Junior Beirut*, docked at the Christian port of Junieh, north of Beirut on 21 May and took aboard APCs and other military vehicles. Informed speculation was that the ship's destination was Israel, where the Lebanese Forces' equipment would be safeguarded. The movement was just in time: the Damascus Treaty was signed next day and it might then have been impossible for the shipment to have got away.[4]

The political significance of this minor operation is that it signals the Lebanese Christians' opposition to the Treaty. As they saw it, Syria had won what it had been working for throughout the 16-year war in Lebanon. President Assad combined the treaty ceremony with the formal opening of his new palace in the mountains of Qassioun, 24 miles from the Lebanese border. Assad has often said that Lebanon and Syria are united 'geographically and

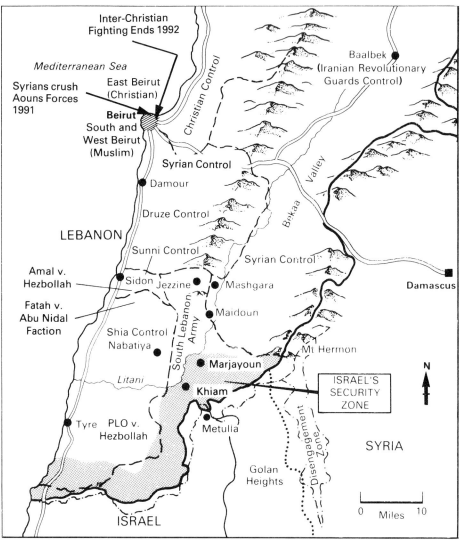

Lebanese regions of conflict as Syria extends control

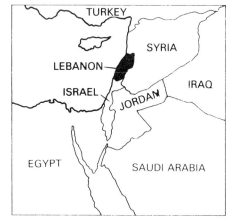

historically and separated only by politics'. At his palace, he called the new relationship 'the work of God'.⁵

Armed resistance to the take-over of Lebanon is unlikely at present. After two years of setbacks and enormous self-inflicted wounds caused by the inter-Christian fighting, the Maronites need time to gather their strength and recover their morale. Also, the community still has divisions. While Sami Geagea rejects the Lebanon–Syria treaty, his second-in-command, Karim Pakradouni, as a pan-Arabist, has a more tolerant approach to Syria. Between them these two men are adopting a pragmatic wait-and-see policy.

Israel and Lebanon

Israel's defence minister, Moshe Arens, has said: 'The *de facto* annexation of Lebanon by Syria is only slightly more subtle than the takeover by Iraq of Kuwait.' A former defence minister, Yitzhak Rabin, warned Syria against altering the *status quo*. He specified deployment of Syrian troops south of their May 1991 lines, interference with the Lebanese army and the introduction of surface-to-air missiles into the area, all of which Israel would view seriously.

In June, the Israeli air force spent three consecutive days attacking a number of targets in southern Lebanon; they included a *Fatah* command post, bases owned by the Popular Front for the Liberation of Palestine (PFLP) and by the Popular Democratic Front for the Liberation of Palestine (PDFLP), posts belonging to Abu Nidal's *Fatah* Revolutionary Council and Ahmed Jibril's training and command post, all in the vicinity of Sidon.

The raids were a warning to all parties in Lebanon that Israel knew the whereabouts of every base and could hit them with impunity. The message was also there for Syria. The director of the Israeli government press office, Yossi Olmert, said: 'If, as I believe, Syria is Israel's number one strategic enemy, then it is decidedly bad for Israel to have this enemy dominating another country that borders Israel.' He was, of course, referring to Lebanon.

Following the Damascus Treaty, the US government encouraged the Beirut government to extend its control over Lebanon. This encouragement was not aimed at reducing the influence of Syria, Lebanon's overlord, but was a shrewd move to provide the security which Israel wants along its northern border. The South Lebanon Army (SLA), under General Antoine Lahad, was unhappy about losing territory to the national army while Israel, Lahad's backer, had no intention of pulling out of the security zone.

The Lebanese army, having been deployed in the mountains east of Sidon, is supposed to be able to prevent attacks not only against Israel but against Christian enclaves, such as that at Marjayoun, which lies in Israel's security zone. The strategic key to the control of much of southern Lebanon is the town of Jezzine, which clings to the edge of a mountain ravine and lies at the tip of a corridor which runs north from the security zone.

It is the home of 20,000 Lebanese, three quarters of them Christian, and is held by the SLA. Around Jezzine are another 10,000 Christian families who fled their homes east of Sidon when the Palestinian and Muslim militias advanced towards Jezzine in 1985. Before the Christians in Jezzine will accept an extension of government power in their region, they want proof that the new Lebanon will allow them to return to their villages — and protect them there.

The Lebanese government, urged by the Syrians, hints that it might use force to send its troops into Jezzine. This would leave 600 Lebanese government troops who have been in the town since 1985 in an impossible position, since they have co-existed there with the SLA for seven years. Jezzine, which Australian troops liberated from the Vichy French in the summer of 1941, is, after 50 years, again a bone of military contention.

The Israeli army patrols Jezzine and they have 155mm guns dug into the fields south of the town to repel any move by the Lebanese army to take the town by force. General Lahad's SLA has 155mm and 130mm guns near Rihan, further south. Within easy reach of the potential trouble areas, the Israelis keep a large number of M-113 armoured vehicles ready for rapid reaction.

General Lahad has several arguments for holding out against the Lebanese army. He says the army has failed to get hold of many of the PLO's heavy weapons in southern Lebanon; it had not disarmed the pro-Iranian *Hezbollah*; Yasser Arafat could order his troops to break out of the Sidon Camps and overwhelm the Lebanese troops; and *Hezbollah* gunmen continue to attack his positions along the mountain corridor between the Israeli border and Jezzine. If all these reasons were not cogent enough, Lahad claims that without the SLA — backed by Israeli promises of support — the long-suffering Christians of the area have no guarantee of security.[6]

On 17 July Israel made its first show of force after the Beirut government moved troops south to extend its authority and prevent guerrilla raids. Shia Muslim guerrillas ambushed an Israeli unit and killed three soldiers, including two officers. The fighting took place at Kfar Houneh, 15 miles north of Israel and outside the security zone. Within hours, Israeli aircraft bombed *Hezbollah* villages close to the village.

The Lebanese Army versus the PLO

Before the Lebanese army moved east from Sidon into the mountains it had made its most decisive move in 16 years, marking a turning point in modern Lebanese history. I have isolated it here from the activities in the mountains — and indeed it was militarily isolated — to emphasise its significance.

The newly-constituted national army, freshly armed by Syria, took on the powerful PLO forces in west Fatahland. Consisting of regular soldiers from all of Lebanon's religions — Shias, Maronites, Sunnis and Druse — the army was deployed east of Sidon and around the vast Ein el-Hilwe camp.

Yasser Arafat, the PLO chairman, had previously said that there would have to be a pact about the future status of the Palestinians in Lebanon before Lebanese troops were allowed to enter this flank of 'Fatahland', the PLO's state within a state. Arafat largely had himself to blame that he was now in such a predicament. Syria, America's ally during the Iraq war, was intent on crushing Arafat's power in Lebanon. Following the Damascus Treaty, the Lebanese army was ordered to move against the PLO and Syria's 'excuse' for beginning this operation was that Arafat had been an ally of Saddam Hussein.

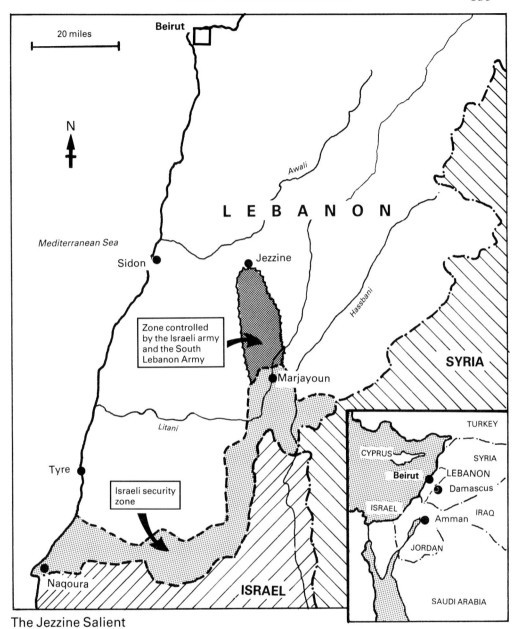

The Jezzine Salient

Even the US was directly involved in the operation. Frederick Veeland, the US Deputy Secretary of State for Near-Eastern and South-Asian affairs, was sent to Beirut to co-ordinate the Lebanese deployment and to ensure that neither the Israelis nor the SLA hindered it, though it was unlikely that they would have wished to do so.

Local Palestinian leaders sought desperately to delay the army's advance by engaging the senior officers in parleys. But the commander of the army's Ninth Brigade gave the same reply to each armed delegation: the army would deploy and take back Lebanese territory whether they liked it or not.

Fighting began with the Palestinians firing shells at the Lebanese army barracks in Sidon. The army, using tanks and artillery to support the infantry, launched a three-prong attack towards the hills that overlook the PLO camps of Ein el-Hilwe and Mieh and Palestinian fighters fled from house to house to avoid the shells. Within three days the army committed 10,000 men to the operation, an indication that the government had no intention of giving in to the PLO's demands for 'dialogue'.[7]

Spokesmen at PLO headquarters in Tunis expressed great bitterness over the events in Lebanon, as well they might, since the PLO would have no base for operations against Israel if it were to lose its military stronghold in Lebanon. The PLO found itself abandoned regionally and internationally and after four days of battle the fighters withdrew into their camps and agreed to surrender heavy and medium weapons. The leaders were particularly bitter that Lebanon's tough military action was actually putting into effect an 'American plot'.

The patriarch of the Maronite church incited hatred against Palestinians by declaring that PLO terrorists had lit the fuse which began the civil war. Many old hatreds surfaced as a result of the Lebanese army's forthright attack.

Israel was pleased at the humbling of the PLO but not particularly relieved and spokesmen said that PLO terrorist attacks might still continue only now 'under the auspices of the Lebanese army'. But nobody who saw the fighting, in which about 300 people were killed or wounded, most of them Palestinians, had any doubt that the Lebanese army meant to impose the government's will.

Michael Murr, Lebanon's Defence Minister, told his soldiers: 'Your guns are pointing the dawn of lasting peace in Lebanon. Go forward — the whole nation is behind you.' Another government minister, Walid Jumblatt, the Druse leader, had a different point of view. 'There is no spectacle more absurd than that of a soldier on a tank on a mission of domestic security', he said, and asked why the army was fighting the Palestinians and not the Israeli army in southern Lebanon.[8]

The simple answer is that President Assad of Syria is not ready to take on Israel. When he does so, Lebanon, as a government and as a nation, will have no say in the matter. There can be no doubt that Assad will want his 'great war' against Israel so that — in the event that he is victorious — history will recognise him as the great leader in Arab and Islamic history.

Hezbollah Attacks: Israeli Reprisals

During the latter half of 1991 and the first two months of 1992, *Hezbollah* and

Palestine terrorists based in the southern Lebanon villages of Kafra and Yater fired rockets against Israeli settlements and raided targets in Israel's 'security zone'. The Israeli defence forces were slow to react because the villages were under the jurisdiction of UNIFIL.

However, on 21 February 1992, Israel sent a column backed by helicopter gunships into southern Lebanon. Israel's prime minister, Yitzhak Shamir, said the army's mission was 'to stop for good' the *Hezbollah* rocket attacks. Israeli gunners fired 200 rounds of 155mm howitzer shells into valleys around Kafra and Tibnin, apparently to block the guerrillas' escape routes. Simultaneously, tanks and armoured personnel carriers, supported by Cobra helicopter gunships, advanced from the security zone and forced their way past the UN units that blocked their way. According to a spokesman for UNIFIL, the peacekeepers suffered six casualties from Israeli fire and in clashes with retreating guerrillas.

Under pressure from Syria and Lebanon, *Hezbollah* and another Shia Muslim guerrilla group, Amal, agreed to cease their rocket attacks. Hezbollah also agreed to keep armed men out of sight in southern Lebanon. UNIFIL observers reported that 'thousands' of fighters who had been flowing south were returning to bases north of Beirut.

However, on 6 April, pro-Iranian Palestinian gunmen attacked an Israeli army convoy near the village of Adeisse. The raid had been well planned for in the convoy were Major General Yitzhak Mordechai, Colonel Shlomo Hassoun, who commands the Israeli troops in the security zone and General Antoine Lahad, commander of the South Lebanese Army.

By remote control, the raiders set off two bombs at a crossroads only 300 yards from Israel's border fence. In the explosions, or in the fire fight which ensued, two Israeli soldiers were killed and five wounded. Three Palestinians, members of the Islamic Jihad movement, were also killed. The Israeli party had been returning from a goodwill visit led by General Mordechai to Shia Muslim villagers in Adeisse at the end of the Islamic holy month of Ramadan. The attack was almost certainly a reprisal for the Israeli assassination of Sheikh Abbas Mussawi, leader of *Hezbollah*, in the same area in February.

In mid-1992, about 1,000 Israeli troops remained in the Security Zone. This buffer strip, ten miles deep, that runs from Mount Hermon in the east to the Mediterranean in the west, has certainly proved essential to the protection of northern Israel. However, the *Katyusha* rockets fired by *Hezbollah*, Amal and PLO fighters often land in the zone itself, inflicting casualties on the Lebanese residents there.

Most patrols that check roads for guerrilla car bombs and other traps each morning, and guard against terrorist infiltrations at night, are from the South Lebanese Army, which is in alliance with the Israeli Defence Forces.

Hezbollah's Endless War

In August 1993 *Hezbollah* guerrillas achieved a victory over the Israeli army that was both tactical and psychological. *Hezbollah*'s 'Voice of Light' radio announced that Israeli troops had suffered heavy casualties from an exploding mine near the village of Shifin, southern Lebanon. This was untrue but the Israeli command suspected that *Hezbollah* had botched an attack against them and had claimed a triumph nonetheless, as so often in the past. They waited for two days before sending in a strong patrol, led by tanks, to search for explosives.

Hezbollah knew that the Israeli army would fall for their trap and as the troops approached the mine they set it off, killing nine Israeli soldiers. They announced that this marked the 24th anniversary of the arson attack by an Australian Christian, on Jerusalem's al-Aqsa mosque.

Syria, which had not been consulted about the mine ambush, cut off the supply of *Katyusha* rocket-launchers to *Hezbollah*, though not the rockets. In Beirut, *Hezbollah* fanatics demonstrated against the Lebanese government for its lack of support in the fight against Israel and nine *Hezbollah* men were killed. Sheikh Nasrallah, leader of *Hezbollah*, demanded the resignation of the government and further demonstrations occurred. This incident took place on the day that Yasser Arafat and Yitzhak Rabin shook hands in Washington.

Nasrallah told the British journalist, Robert Fisk: 'If all the world were to recognise Israel, we never will. Israel has been established on land that has been seized from others.' Even a compromise truce could not take place until Israel withdrew its troops from the security zone and returned them to the international border, Nasrallah said. Israel responded by saying that it would not retreat from southern Lebanon until six months after *Hezbollah* laid down its arms. The Clinton administration has supported the Israeli stance by telling the Lebanese government that it will allow no American to visit Lebanon until *Hezbollah* is disarmed.

Even Syria would like *Hezbollah* out of the way - though no government minister would dare say it - in order to move a step closer to an accord with Israel. However, Iran, *Hezbollah's* principal sponsor, would resist every attempt to emasculate its creation.

In the meantime, *Hezbollah* and *Fatah* are in conflict in Lebanon because *Hezbollah* regards *Fatah* as a supporter of peace between Israelis and Palestinians.

In the first half of 1994, *Hezbollah* fired rockets at targets in settlements in northern Israel, thus triggering an Israeli response; aircraft attacked *Hezbollah* bases on several occasions. *Hezbollah* also attacked troops of the South Lebanese Army.

The Lebanese army, slowly rebuilding its strength after the destruction caused by years of civil war, has regained much of its authority but it dares not confront *Hezbollah* openly. To do so would anger both the Iranians and the Syrians.

References
1. According to Maronite leaders in Beirut, Crocker said as much.
2. The French press picked up this comment and widely quoted it.
3. From a French intelligence source.
4. The ship's loading and sailing was witnessed by a British diplomat.

5. According to a Syrian communiqué issued in Damascus and Beirut. In East Beirut, a Christian centre, Assad's words were changed to 'the work of the devil'.
6. As outlined by Robert Fisk in *The Independent*, London, 18 July 1991. Fisk commented that it was highly unlikely that the PLO could overwhelm the Lebanese army.
7. According to a French diplomat in Beirut: 'The Lebanese have learnt from bitter experience that nobody wins a dialogue with the PLO. PLO spokesmen are masters of bluff and deceit and they know every possible delaying tactic. Trying to negotiate with the PLO was largely responsible for landing Lebanon in the mess it has been in since 1975.'
8. Quoted by Robert Fisk, *op. cit.*

Morocco — Polisario War

STRAINS ON THE REFERENDUM

Background Summary

In 1976 Spain withdrew from its former colony of Western Sahara, and Morocco and Mauritania partitioned the country. They had not reckoned with the Sahrawis' resistance movement, the Popular Front for the Liberation of Saguia el-Jamra and Rio de Oro (Polisario). After Polisario defeated a Mauritanian force, Morocco took the opportunity to claim the entire area.

Unable to occupy the country by force and in the face of Polisario harassment, in 1980 the Moroccan army built the Hassan Wall, a remarkable system of fortified defences, to keep out Polisario and the Sahrawis generally. However, the Sahrawis found refuge in Algeria from where their leaders declared the independence of the Sahrawi Arab Democratic Republic (SADR). The Organisation of African Unity (OAU) and a total of 70 countries had recognised SADR by 1990 but the UN has not recognised it because of Moroccan and Arab lobbying.

The Moroccan army has never been able seriously to restrict Polisario's activities, despite a Hassan Wall garrison of 110,000 and advanced surveillance systems. Some 80,000 troops are in the area. When Polisario has reduced its raiding activities it has been at the request of its Algerian hosts. Polisario actually suspended hostilities for nine months while testing the sincerity of Morocco about a proposed referendum in Western Sahara. King Hassan II refused to negotiate on SADR's demands for independence and Polisario raiders breached 10 miles of the wall in a massive assault in 1989.

Several encounters took place during 1990 and Morocco extended the length of the Hassan Wall, while spending 72 per cent of the national budget on development in Western Sahara. After many difficulties, UN officials, notably the Secretary-General, Perez de Cuellar, and his special representative, Johannes Manz, produced a manual for a referendum on the Western Sahara issue. It was based on the UN's experience in Namibia, where negotiations had brought about a successful peace.

The War in 1991–92

On 15th anniversary of the proclamation of SADR on 27 February 1991 took on a special meaning after Iraq's invasion of Kuwait in 1990 focused international attention on the sovereignty of nation states. While the Sahrawis, now 176,000 of them, live in refugee tents, the Moroccan King and government still cannot agree to the terms of the UN peace plan for a referendum.

The delay in ending the conflict lies in the implementation of a UN Resolution to enable 'all Saharans originating in the territory to exercise their inalienable right to self-determination'. The Sahrawi people have always recognised the resolution but, through the presence of what they describe as the 'Moroccan aggressor', they have considered it necessary and just to fight for independence and peace.

Whether King Hassan's peace process will involve the withdrawal of the Moroccan administration and army from the territory is the main obstacle to what Polisario feels is the only free and fair way of reaching agreement. Before a referendum is held, Polisario wants the installation of a UN/OAU administration during the transition period, the release of all Sahrawi prisoners detained by Morocco and the return of all refugees to the Western Sahara.

'Fairness' is not one of King Hassan's attributes.[1] He has been a tyrant within Morocco, as was shown in January 1991 by his government's brutal handling of riots in Fez over poor living conditions, and by a dreadful human rights record. In December 1990 Amnesty International reported to the UN that several hundred people had 'disappeared' after being arrested by the Moroccan security forces since 1975. Many of the victims came from the south of Morocco and from Western Sahara, and were suspected of supporting Polisario.

In 1990 King Hassan cancelled a year of Moroccan cultural events in France. The *L'Année du Maroc* was abruptly called off ostensibly 'because of the Gulf War', as Moroccan officials said. The Algerians and Polisario leaders believe that King Hassan was fearful that Sahrawis and exiled Moroccans would use the cultural events as a focus for their protests.

The UN Secretary-General's second report on the Western Sahara referendum had been expected in August 1990 but it was delayed because of the Iraq war. On 17 May 1991 the UN General Assembly approved Resolution 690, which proposed a budget of $200 million for the referendum programme. It also specified that the Moroccan forces would be confined to certain areas and not exceed 65,000. Finally, it laid out a timetable for the peace plan; the ceasefire date, or D-Day, would be 6 September. Meanwhile the UN would set up Minurso (*Mission d'observation des Nations Unies pour le referendum au Sahara Occidentale*). Minurso would have three units — a civilian force of UN civil servants, a military force of 1,695 soldiers and 300 police. It would comprise representatives from 34 nations.

The repatriation of the Sahrawi refugees from their camps near Tindouf, southern Algeria, to the Western Sahara would begin no later than 11 weeks after the ceasefire date and would be complete within six weeks, while the whole referendum operation must be finished within 36 weeks.

The SADR and Polisario leaders anticipated dirty tricks. In particular, they were suspicious of the presence of thousands of immigrant Moroccans from the north. Usually without work, these Moroccans had been brought in to the towns of El Ayoun and Smara to take advantage of the high salaries, low unemployment and cheap prices. Every aspect of life is subsidised by the Moroccan government to encourage immigration to the region. Whether these immigrants, nearly all of whom are sympathetic to the Moroccan claim to Western Sahara, should be able to vote was a major problem in the holding of

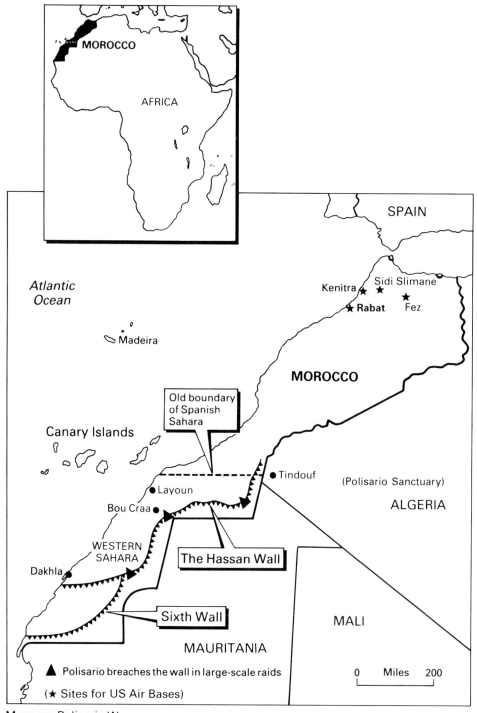

Morocco-Polisario War

the referendum.

In mid-1991 King Hassan was unlikely to compromise about anything, since so much was running his way. Polisario attacks had become less frequent and less threatening. With the wavering effectiveness of the guerrillas, the cut in Libyan aid to Polisario, the solid economic and social entrenchment of Morocco in the Western Sahara and the erosion of hope among the Sahrawi people, Hassan had no incentive to respond to UN pleas to 'show flexibility'. Nevertheless, he called an end to hostilities early in July.

On 6 August the entire issue of the referendum was upset when the Moroccan air force attacked Tfariti in the Western Sahara and repeated the attack next day. A Moroccan government statement in Rabat, carried by the official news agency *Map*, said: 'The Royal Armed Forces have conducted and are conducting sweeps and clean-ups in no man's land.' This referred to the area between the Moroccan defence lines and the Western Saharan borders. The report made no mention of Tfariti.

The reason for the attack, according to the government, was the Polisario guerrillas had infiltrated into no-man's-land with the 'clearly defined mission' to commit terrorist acts inside Moroccan Sahara. In the phraseology of the report, 'the guerrillas' aim was to perturb and delay preparations for and the peaceful conduct of the self-determination referendum'.

Polisario's counter-statement was direct. Without provocation, Moroccan aircraft had attacked the guerrilla-held oasis, which lies about 10 miles north of the Mauritanian border in the north-eastern corner of the disputed territory. According to Polisario, it is about 60 miles outside the Hassan Wall. Only 24 hours earlier, Polisario reported, 15 Moroccan aircraft had carried out a 'massive attack' on the post. They had shot down one plane but the report gave no details about casualties.

Ibrahim Hakim, a SADR roving ambassador, said after the first attack that he hoped it was an isolated incident but after the second assault SADR's information minister, Bur Lahlou, said: 'The credibility of the United Nations and its Security Council is being challenged by Moroccan adventurism'.

The Polisario leader, Muhammad Abdulazis, protested to the UN about the attacks. 'I am convinced that if this deliberate de-railing is not promptly ended, the chances of peace will be destroyed', he told the Secretary-General.

Despite the setback, the first UN personnel duly arrived in Western Sahara on 10 August and began to make preparations for the referendum. None of the many UN officials, Algerian ministers and diplomats in Morocco to whom I spoke believe the referendum will end the conflict. The consensus is that either the whole affair would be seen to be fraudulent in Morocco's favour or that, if the vote went against Morocco, King Hassan would not accept it and would find a way of breaching the result.

Low-Level Conflict 1993-94

Because of violent upheavals in areas regarded by the Western nations as 'more important', and because of an assumption that the referendum had brought peace to Morocco, little was heard of Polisario's fight for freedom during 1993-94. In fact, it has continued but at a lower level.

As was widely expected, the referendum ballot showed a majority in favour of Western Sahara joining Morocco, but the vote was suspect. UN observers reported numerous irregularities by the Moroccans, including coercion.

The Sahrawis themselves observed the proceedings with eagle eyes and expressed many complaints to the UN staff. Neither side accepts that the war has ended and in April 1994 the garrison of the Hassan Wall had not been reduced. The Sahrawi leadership has been undecided about the future but to emphasise that they still want independence they send Polisario units into Moroccan territory on raids aimed more at civil installations than army targets. Observers predict that hostilities will simmer for years but some believe that if the Moroccan occupation eventually comes to be seen as benign, the war might then end.

References

1. The French government was so disturbed about King Hassan's human rights record that it commissioned a secret psychiatric study of the monarch. The specialists who carried out this work suggest that Hassan has become a megalomaniac. This condition, they say, is the result of his fears of assassination or overthrow; he had survived three known assassination attempts and two attempted coups. Hassan's strong-arm tactics, the psychiatrists report, are his reaction to the threats against him. Only by completely subjugating the government and people to his will can he feel safe. He forms the policy for every government department and he makes all major decisions. I understand that the report says that Hassan's hatred for Polisario is so intense as to be pathological.

Mozambique Guerrilla War

MOVING VIOLENTLY TOWARDS PEACE

Background Summary

Mozambique became independent from Portugal in 1975 and the Mozambique Liberation Front (Frelimo) won control of the country in the turmoil that followed. Through its economic and social policies, it brought the new nation close to ruin. The country's destruction was complete when the Mozambique National Resistance (MNR, better known as Renamo) came into being as a result of scheming by the Rhodesian Central Intelligence Organisation. The Rhodesians, led at the time by Ian Smith, planned to use Renamo to sabotage any assistance which Frelimo might give to Zimbabwean independence movements.

South Africa entered the conflict in 1982 to ensure that Mozambique would not become a threat to its security. Renamo was aided not only by South Africa but by Portugal, Morocco, Saudi Arabia and Zaire. Frelimo, under President Samora Machel, depended on Chinese and Soviet aid. Killed in a plane crash in October 1986, Machel was succeeded by Joaquim Chissano.

By then the war had become one of the world's most vicious. US Assistant Secretary of State, Roy Stacey, described it as 'a brutal holocaust, a terror war against Mozambican civilians'.

In 1989 foreign aid to both sides declined. President F.W. de Klerk of South Africa, Robert Mugabe of Zimbabwe and Daniel arap Moi of Kenya urged the parties towards peace talks, but the ideological differences were great. Frelimo insists on a one-party state in the Marxist–Leninist style, while Renamo demands a constituent assembly and a multi-party system.

The War in 1990

From mid-year, many thousands of people who had been refugees in neighbouring countries began to return home as *recuperados*. Their homes and settlements were in ruins and aid agencies reported that it would take about two years before they became habitable.

Frelimo strategy was to establish pockets of government control so that civilians could resettle, but vigilance was needed to ensure that Renamo marauders did not counter-attack, slaughter the populace and deter settlement for years. By October, the Frelimo forces were achieving military successes against Renamo in the provinces of Zambezia, Tete, Manica and Sofala. Trained by Soviet instructors and highly disciplined, the Red Berets practically eliminated attacks on transport moving along the highways.

Simultaneously, a 28-year-old traditional healer, Manuel Antonio, brought a form of 'holy war' to Mozambique when he organised a militia battalion known as *Naprama*, a Macua word for 'irresistible force'. Antonio claimed that anybody whom he 'vaccinated' could not be harmed by bullets or shells. His men were so inspired that, armed with nothing more than spears and machetes, *Naprama* overran Renamo positions in Zambezia province which the army had been unable to dislodge.

When the Communist countries of Eastern Europe collapsed, their help for African Communist states dwindled and in some cases ended. Mozambique was so badly affected that President Chissano had no option but to listen to the urging of the Western powers to talk with the rebels. In an attempt to win foreign support, the government presented a new draft constitution which guaranteed so many democratic rights that it was praised by President Bush.

The War in 1991–92

Army discipline broke down early in 1991 because of the lack of food. Partly because a severe drought caused a 90 per cent loss to the maize crops, food was scarce everywhere. Throughout the 1980s there had been incidents of army theft but in February 1991 troops escorting a convoy of relief supplies in Zambezia province unloaded it before reaching their destination. In February, troops shot dead six civilians as a relief ship arrived in Moma, on the border with Nampula province.

When regular soldiers and militiamen repeatedly stole food intended for war victims, one international relief agency suspended an airlift to the town of Murrua. It is difficult not to feel some sympathy for the troops, who sometimes get no issue of rations for three months. When food finally arrives they are told not to touch it.

Even as a new round of peace talks was beginning in Rome on 25 April 1991, the army was facing unrest in the ranks. Yet a disciplined national army was essential if any reforms were to be put in place. Italian mediators made this point, but they were hopeful that a nationwide ceasefire could be signed between Frelimo and Renamo.

The most important points for discussion at the peace talks concerned means of integrating the two warring armies, a date for the general elections and whether Zimbabwe could be induced to withdraw its 7,000 troops. This force was guarding two strategic transport routes through Mozambique which link landlocked southern Africa to the Indian Ocean.

President Chissano's new constitution was bridging the formerly wide political differences. The key clauses concerned multi-party democracy, elections by secret ballot and respect for private property. In short, Marxism was dead in Mozambique and, by April, five opposition parties were in their infancy.

Foreign observers could see that the key to any ceasefire was the demobilisation of large numbers of the troops and the integration of those who were left into one professional army. Renamo's leader, Alfonso Dhlakama, was reluctant to order his men to lay down their arms since, understandably, he did not trust

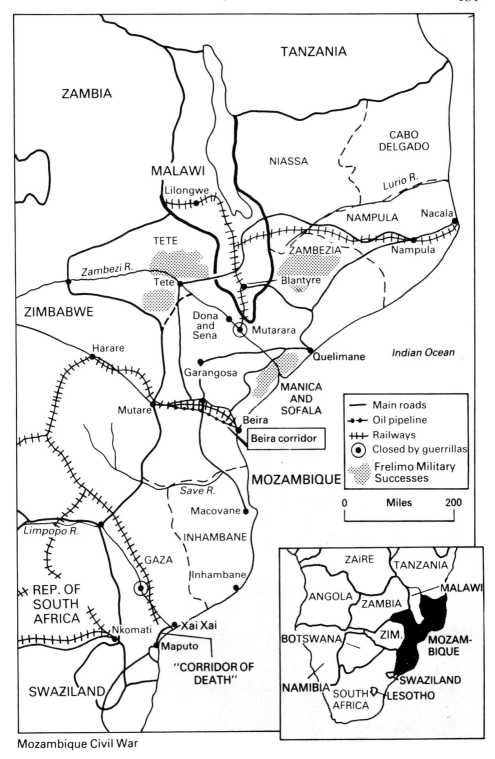

Mozambique Civil War

the government. Chissano and his ministers were also cautious because of the rebels' record of breaking agreements.

Swiss government officials were helping the Mozambique Ministry of Finance to evolve a plan to provide jobs for officers and men listed to be discharged, but the troops had no confidence in any plan. Sixteen years after independence from Portugal, thousands of veterans were still living in poverty. In April, many observers were blaming the erosion of army discipline on the anticipated military changes that peace would bring. 'War brought misery to civilians', said a Soviet diplomat, 'but for the army it brought a livelihood and security.'

Whatever hopes the peacemakers had for the success of the Rome talks — the sixth in four years — were dashed when Renamo began an offensive in southern Mozambique. The movement had been increasing the pressure on the government for further concessions for some months, with attacks on homesteads on Maputo's outskirts, sabotage of power lines and attacks on the railroad to South Africa and National Highway No 1 in the north.

On 8 May an atrocity occurred which seemed to prove the government's contention that Renamo could never be trusted. A Renamo platoon of 17, at least 12 of them young boy soldiers, raided the village of Ndavela, battered down the door of Laice Rodriguez and demanded to know the whereabouts of local government officials, militiamen — and the town's grocery stores. With Rodriguez and his wife, Olinda, and another man as their captives, the raiders went from house to house looking for food and *tontonto*, a heady home-brew. In the process they killed an unarmed militiaman and beat senseless a man who refused to give them *tontonto*.

With their prisoners, the Renamo squad set off to march to Michafutene, a town in the rebel zone. When two elderly men complained that they could not keep up, the leader ordered several of the boys to chop off their ears and lips and then release them.

The next day they cut off the ears of the other victims and made deep knife wounds across the bridge of their noses. The leader said: 'Go and show your ears to President Chissano', and the gang prodded them into the bush. The victims finished up in hospital in Maputo. Mr. and Mrs. Rodriguez said that the boys were aged about eight or nine and that the gang kept five other boys they had picked up in the raid. They were to be trained as soldiers in Renamo's army.

Alfonso Dhlakama repeatedly assures journalists[1] and foreign officials that his army does not mutilate civilians: 'Of course, this can happen to Frelimo soldiers', he said in one interview. However, there are hundreds of documented cases of mutilation of civilians, notably in Sofala, Manice and Tete provinces.

Critics of the peace plans use such incidents to stress that while Renamo as an organisation might agree to a ceasefire, the countryside is full of brutal, lawless marauders who prevent a return to normal life.

The Triumph of Spiritual Warfare

As foreigners grappled at a distance with Mozambique's problems during 1991, traditional leaders and healers *(curandeiros)* assumed much greater importance. Manuel Antonio remains the best known of them because he has been interviewed by foreign journalists but many others are significant factors of social control and rehabilitation. Frelimo has viewed these men — and some women — as enemies but in 1990 President Chissano ordered his officials to cooperate with them, if only because the traditional leaders opposed Renamo. Several of them ran orderly communities behind enemy lines.

The most powerful traditional chiefs are *mambos* but the more numerous *curandeiros* are highly regarded. The authorities have learned that they themselves can get nothing done unless the chiefs back them. The government has even approved the formation of associations of *curandeiros*.

Some spiritual mediums have been remarkably successful in keeping fighting out of areas they control. The medium Mungoi kept inviolate a region of about 20 square miles in Manjacaze, a district of Gaza province where conflict and violence has always been rife. Even the most vicious Renamo thugs are afraid of Mungoi, though foreign inquirers have been unable to find out why this man is so potent and powerful.

With his peasant militia, armed only with token weapons against Renamo's AK47 assault rifles, Manuel Antonio continues to be successful. Remarkably, he has yet to fight a battle. Renamo units melt away in the face of his 'warriors' whose most effective weapon is the spell with which Antonio protects them. He has great if undemonstrative authority in Zambezia and Nampula provinces.

There is, however, another side to the communities in which Antonio and others exert their age-old mystic power. It is the ignorance of modern medicine and distrust of its practitioners. In Zambezia, Nampula and other provinces, whole communities have chased away vaccination teams sent in by the Mozambique Ministry of Health and UN health agencies. They fear that the needle and syringe will be used to drain the people's blood. Native Mozambicans have for centuries feared blood-sucking. Mozambique is rife with disease after 16 years of warfare but medical teams cannot induce the people to accept inoculation against whooping cough, tetanus and measles.

By the end of 1991 peace was coming slowly to Mozambique. The political peacemakers believed it would arrive in 1992 but the more realistic UN officials and the workers from foreign aid agencies, who spend much of their time in the bush with the people, said that even if treaties were signed a practical workable peace would not be in place for another five or six years. With eight-year-old boys being trained as fighters and taught to cut off the ears of 'enemies', the dangers of war were ever present.

Mozambique's Press-Ganged Army

An extraordinary situation developed in Mozambique at the end of 1991 and continued into 1992. While the government was negotiating with international donors for funds to demobilise troops the army was actually recruiting men and boys by the thousand.

At first the campaign was directed against draft-dodgers. They were offered an amnesty if they would report to recruitment centres by a certain date. The ploy was a total failure and the army was ordered to switch to hunting down draft-dodgers. This proved difficult so army patrols waylay young men in the streets or take them off buses. Recruits have been stripped of their shirts and shoes, roped together and taken away in military vehicles. Recruiters are not interested in the legal status of men arriving at the barracks; anybody who appears to be of military age is taken into the army.

Abuses are flagrant. Teachers and pupils have been dragged out of schools. According to allegations, dozens of press-ganged recruits were locked in a container overnight at Maputo airport, where they suffocated. The diplomatic corps takes this story seriously but the Defence Ministry issued a formal denial.

The number of recruits seized by press-gangs is a 'military secret', but Radio Mozambique's investigative programme, *Onda Matinal*, suggested in April 1992 that it was 70,000.

The War in 1993-94

The end of the Cold War meant that the former Soviet Union and its East European satellites could no longer support Marxist/Communist regimes in various parts of the world. The end of this military aid had a disastrous effect in Mozambique where Frelimo found itself politically isolated, except from China. The Chinese had always supported Frelimo and after Soviet and East German aid dried up - there was no longer an East Germany - Frelimo turned to Beijing for increased support. The need was desperate because Saudi Arabia's aid to Renamo had increased.

The Chinese government hesitated, apparently because it did not want to anger the US government, first of George Bush and then of Bill Clinton. At least, this is the view put forward by foreign observers in Maputo. Finally, the Chinese agreed to continue and increase aid to Frelimo. Again according to diplomats in Maputo, supported by others in Beijing, the Chinese are using Mozambique as a proving ground for weapons, ammunition and all kinds of military equipment. They are able to get away with this because Mozambique has faded from the world's attention. The focus of interest since mid-1993 has been South Africa, thus ensuring China of a free run in Mozambique. Renamo has reported that its forces have captured new weapons of Chinese manufacture as well as much personal military equipment made in China.

China has also been supplying Frelimo with food stocks and, according to some sources, with cash. Frelimo instructors are being trained in China - replacing the training they had previously been given by the Soviet army. Some Chinese military advisers have been seen in Mozambique but none has been captured by Renamo.

China has another reason to help Frelimo. The movement wants a one-party Communist state on the Chinese model, which is an adaptation of the Marxist/Leninist model. China has fewer opportunities than ever to experiment abroad with its political doctrine and Mozambique must seem especially attractive. In the meantime, the people of Mozambique continue to suffer and Western aid has decreased because of the great demands from other countries, such as Somalia.

References

1. Few Western journalists cover Mozambique on a regular basis. Other than Portuguese reporters, perhaps the only European based in Mozambique is Karl Maier, who writes for *The Independent*, London. Maier is reputed to know more about Mozambique than any other Westerner, a knowledge gathered largely through his courage in venturing into parts of the country where companies of élite troops might fear to travel.

Northern Ireland Terrorist War

THE LIBYAN CONNECTION

Background Summary

The rival communities of Northern Ireland have been in confrontation for centuries. Sinn Fein, whose members are Roman Catholic, seek the union of Northern Ireland's six counties with Eire, the Irish Republic. The majority Protestant or Loyalist community insists on the Six Provinces remaining part of the United Kingdom. In these opposing causes thousands of people, most of them non-military, have been killed.

Sinn Fein split in 1969 and since then its military arm has been the Provisional Irish Republican Army, generally known as the Provisionals or Provos. In practice, Sinn Fein and the IRA have the same command. The IRA wages a terrorist war against the Protestants' 'defence' groups, some of which are also terrorists.

The present war began in 1969 when the British army was sent to Ulster to protect both the Catholic and Protestant communities and to fight the terrorists. IRA propaganda represents the British army as an alien 'occupying' army, while the Royal Ulster Constabulary (RUC) and the Ulster Defence Regiment (UDR) are described as traitorous collaborators with the British. Most of the Roman Catholic community appears to believe this.

The army had 10,200 men in Northern Ireland in 1987. This military presence restricted the IRA's operations in the province so it sought targets in Germany, Gibraltar and elsewhere. In mid-1988 it carried out a campaign of destruction against commercial premises and offices in Northern Ireland.

In 1989 the IRA's strategy was 'diversification' — numerous indiscriminate attacks on soft targets. These went on into 1990, both in Northern Ireland and abroad. One of its most 'successful' operations was the murder, by means of an under-car bomb, of Ian Gow MP, a Conservative spokesman on Northern Ireland.

The aim of the IRA was — and remains — to bring Northern Ireland terrorist conditions to mainland Britain and more incidents took place in 1990 than in any year since 1974. The IRA regards the 1990 'British campaign' — as it calls the attacks — as its greatest success. On 19 September the Prime Minister, John Major used the word 'war' when describing IRA activities and the IRA took this as a gratifying sign that its own terminology was being accepted in British circles.

IRA violence reached a new level of depravity in October 1990 when terrorists first used proxy-bombers. They kidnapped men uninvolved in the fighting, strapped them into vehicles loaded with explosives and forced them to drive into military checkpoints, where the bombs were detonated by remote control.

The security forces' victories were infrequent but significant. On 9 October 1990 the SAS carried out an operation in which five IRA terrorists were ambushed near Loughall, Co. Armagh, and two of them, important IRA killers, were shot dead.

The War in 1991–92

Throughout the earlier part of 1991 the Northern Ireland Secretary, Mr. Peter Brooke, strove to find a political solution to the problems of the troubled land. All-party round-table talks collapsed in July when the British and Irish governments refused to cancel a pre-arranged meeting of the Anglo–Irish Conference. Violence followed and by mid-November 86 people had been killed during 1991. Two of them were members of the IRA on 'active service' in Britain. They were assembling a bomb in St. Albans when it exploded.

As the violence continued, the head of the RUC, Sir Hugh Annesley, raised the prospect of internment without trial. His political superior, Brooke, pointedly declined to rule out internment, which he described as 'a potential weapon of surprise for the security forces'. The carnage of the second week of November, when nine people died in Northern Ireland warfare, brought the controversial policy of internment even further into public debate.

A decision to reintroduce internment would carry high risks. The British government and the security forces remember the failure of internment in 1971–75, when nearly 1,000 suspects were in custody, yet violence increased. The IRA has always hoped for the reintroduction of internment, knowing that nothing could more effectively bind the Roman Catholic community together against the British.

In the 1970s, unreliable intelligence led to many innocent people being interned for long periods. The British government was criticised for permitting 'inhuman and degrading' interrogation techniques which contravened the European Convention on Human Rights.

Late in 1991 some security chiefs insisted that they had learned many lessons from the 1970s internment period. Internment would not now be aimed primarily at the Catholic community and with improved intelligence security chiefs were confident that they could pick out the 300 to 400 'instigators' of violence.

Locked in a Time-Warp

For outside observers, the Protestants and Catholics of Northern Ireland seem to be trapped in a strange time-warp. They chant obsolete slogans and are savagely excited by ancient demons. For them, political realities have not changed in 20 years, yet in all civilised countries the Northern Ireland type incessant violence seems crude and old-fashioned.

Internment might have a palliative effect for a short time but fresh and eager young men of violence would quickly replace the locked-up veterans. In the long term, peace in Northern Ireland can only be achieved through a political

158 THE WORLD IN CONFLICT

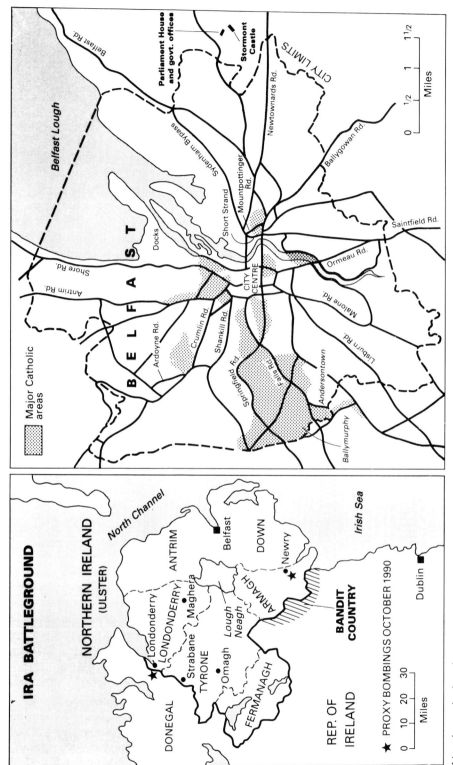

Northern Ireland

solution that deals with the Catholics' deep historical sense of inequality. This would not mean incorporating Northern Ireland into Eire, as Protestants fear, but providing the Catholics with a large degree of assurance that they will never again be subjected to Protestant discrimination.

Protestants distrust the British government's pledge not to change the province's status without a clear majority, but most Dublin politicians and senior civil servants have grave doubts about the viability of a united Ireland, despite their rhetoric. They would be absorbing a nest of virulent malcontents who would turn their endless anger on the Dublin government and bring violence to the entire island of Ireland.

The 1985 Anglo–Irish Agreement, which provides Eire with a consultative voice in the affairs of the province, allayed some Catholic fears and may have averted a slide into even greater alienation, political instability and warfare. But the Protestant Unionists continue to oppose bitterly any role for the Eire government in Northern Ireland's affairs. This was the key factor in the collapse of the Brooke initiative, which was intended to establish a new political order in the province.

The Libyan Connection

The IRA's strategy has been largely based on its arms and explosive links with Libya. Intelligence services had known of this connection since the 1970s but in October 1987, with the seizure of the IRA gun-running ship *Eksund* in French waters, the full extent of Colonel Gaddafi's help for the IRA became known. The *Eksund* was carrying 150 tons of arms and in March 1991 the five-man crew, three of them IRA terrorists, were sentenced to prison terms of between five and seven years.

However, four other shipments got through, giving the IRA an estimated 110 tons of modern weaponry. The list includes tons of explosives, close to 1,000 rifles, hundreds of pistols and grenades, anti-aircraft guns, heavy machine-guns, SAM-7 ground-to-air missiles and flame-throwers. Some of this stock has been recovered by the security forces but in 1992 the bulk remains in IRA hands.

Much has been made of the SAM missile as the IRA 'secret weapon' but Semtex explosive has made the biggest difference to the IRA's campaigns since the mid-1980s. A little Semtex has a devastating effect. The drogue bomb is an example of its use. A pound of explosive inside a tin can with a throwing handle, though it looks crude, is a serious threat to armoured vehicles. To meet the danger, the army has employed many more foot patrols and ordered Land Rovers to travel in threes, not in twos as before. Urgent and expensive research has produced new forms of armour for specially reinforced Land Rovers.

The IRA also uses Semtex in the mortar bombs it fires at army and police bases. Half-a-pound of the explosive is used in each of the bombs, which are made in illegal mortar factories in Eire and smuggled north. Such a bomb was used in the mortar attack on Downing Street in 1991.

The Libyan-supplied Semtex has greatly helped the IRA with its use of under-car booby traps. As a mere two pounds of Semtex is needed to destroy a vehicle and kill its passengers, the entire device can be fitted into a plastic lunchbox and attached by magnet beneath a vehicle. Older, less potent explosive might require several pounds and a much larger container for the same destructive effect. A Semtex under-car bomb killed six soldiers in Lisburn in June 1988. That year also, Semtex ripped apart an army bus at Ballygawley, Co. Tyrone, killing eight soldiers and wounding 28 others.

The IRA has plenty of Semtex and this, together with the arms, is sufficient to keep the terrorists active for the rest of the century. The Libyan link, which even now may not be entirely severed, has given the IRA its cutting edge as a military force.

While SAM-7s have not been used, several army helicopters have been hit by heavy machine-gun fire. RPG-7 rocket-launchers have also been used against security forces' bases and armoured vehicles. Troops on patrol in Belfast found a flame-thrower ready for use; these weapons use a type of napalm and can project a jet of flame 80 yards.

The IRA and Drugs

In October 1991 about 30 alleged drug dealers were told by the IRA to get out of Northern Ireland or 'suffer the consequences'. Four of them had already been wounded in punishment shootings. The IRA's ultimatum was part of an effort to assert its authority in working-class Catholic areas and to try to distance itself from other paramilitary groups involved in drug dealing.

The four men shot in Catholic West Belfast suffered leg and arm injuries of a particularly severe nature, according to police sources. In January 1989 IRA gunmen carried out other punishment shootings in a purge of petty thieves and joyriders whom the IRA suspected were acting as police informers.

Senior police officers admit that the IRA does not deal in drugs, but there is evidence that the organisation has demanded protection money from dealers. The IRA is not necessarily acting on principle by not engaging in the drugs traffic. The simple economic fact is that, in the past at least, it has received all the capital it needs from Irish communities in the US.

Attack in London

On 11 January 1992 a briefcase bomb exploded in Central London, about 300 yards from where the Prime Minister, John Major, was due to hold a Cabinet meeting in Downing Street. The security services regarded the bombing as an IRA boast that it could explode bombs at will. A 30-minute warning was given, so that police were able to seal off Whitehall and Downing Street and nobody was hurt.

The explosion was close to the position from where men with a van fired a mortar bomb into the garden of 10 Downing Street while the Cabinet was in session during the Gulf War in February 1991.

On the same day, 48 hours after a new security plan took effect in Northern Ireland, the IRA hit Ballymena, north-west of Belfast, with five bombs. The consequent fires destroyed two stores and damaged others. A car-bomb in Londonderry went off near police HQ.

In a statement, the IRA said that the London bomb was 'one of the inevitable consequences of British interference in Irish affairs'. This is standard IRA language. The Prime Minister said: 'The attack is utterly counter-productive and it will not shake Britain's determination to reach a negotiated settlement.'

Only two days after the London bomb, a large IRA arms cache was uncovered in West Belfast. The haul included nearly 50lbs of Semtex, 50 incendiary bombs, an under-car booby trap, two rifles, a handgun and 400 rounds of ammunition. Also found were silencers, parts of rocket launchers, detonator cord, radio receivers and masks.

The incendiary bombs were particularly worrying, since the IRA has not used them since 1980. In 1978 it destroyed La Mon House Hotel, London, with incendiaries and 12 people perished.

The commander of the police anti-terrorism squad, Commander George Churchill-Coleman, told a press conference on 5 December 1991 that he believed that only about six IRA terrorists were operating on mainland Britain. However, young people being trained as subversives were then sent to Britain to help in the campaign there.

One of the IRA's most effective ploys during 1991 was to warn that its agents had left a bomb on a railway line in the London area. This inevitably led to an intensive operation by the security services, with consequent disruption to London rail traffic. During 1991, the IRA set off a bomb in a waste bin at Victoria Station and one man died in the explosion. It is believed that the IRA command proposes to cause further disruption to transport with hoax warnings of bombs. To bring the nation's capital to a virtual halt in this way is considered a great propaganda victory since such incidents are reported worldwide.

New Terror Tactics

On 10 April 1992, the day after the British national elections, the IRA set off two massive bombs in London. One, consisting of about 200lb of Semtex explosive, exploded in the City of London, killing three passers by and wounding 91. The second bomb exploded a few hours later at Staples Corner, North London. Nobody was hurt but the damage was immense in both cases and was estimated at £1 billion.

Both bombs were contained in white vans and were triggered by timing devices. The IRA said that the loss of life and injuries resulted from 'the British police's failure to act promptly' on the coded warning that its agents had given. The IRA has frequently given this cynical disclaimer of responsibility for death and injury. Commander George Churchill-Coleman, accused the IRA of deliberately misleading police over the location of the bomb.

Security advisors warned the Prime Minister, John Major, and his new government that the IRA had posted a highly trained 'active service' unit to London and that it was led by two veteran bomb experts. In April, Intelligence sources believed that up to 12 IRA terrorists were active on the mainland. Churchill-Coleman said that the terrorists had embarked on a campaign with varying tactics and on a new scale. 'They are bringing the tactics they employ in Northern Ireland to the mainland,' he said. The 'varying tactics' were demonstrated within two days of Churchill-Coleman's warning: An off-duty soldier in civilian clothes was shot in the head by gunmen of Irish National Liberation Army (INLA), an offshoot of the IRA, as he walked through the centre of Derby.

The IRA's successes and the great difficulty of the police in bringing terrorists to justice shows that even better Intelligence is needed. In mid-1992 the security forces were debating the need to revise the roles played by Special Branch, MI5 and Military Intelligence. In November 1991, 2,000 extra troops had been sent to Northern Ireland but a larger force on the ground does not necessarily overcome terrorist activity.

Libya Hands IRA Secrets to Britain

In May 1992 Colonel Gaddafi of Libya realised that some decisive step was necessary in order to improve his relations with the West. The UN had ordered sanctions against Libya to force Gaddafi to surrender two intelligence agents believed to be responsible for bombing Pan Am flight 103 over Lockerbie in December 1988. He offered to reveal details of Libya's dealings with the IRA and early in June his representatives met a British Foreign Office official and gave him much valuable information.

It is now known that the Libyans provided more training and money to the IRA than British Intelligence had believed possible. The arms included at least 10 tonnes of Semtex explosive, hundreds of rifles and pistols, thousands of rounds of ammunition and quantities of detonators and timing devices. Contrary to earlier reports, no surface-to-air missiles were on the list.

Gaddafi admitted giving the IRA many millions of pounds in American dollars, French francs and German marks. During the 1980s the Libyan dictator was the IRA's chief single source of income.

Most importantly, the Libyans gave the names of 20 IRA terrorists whom they had trained in terrorist camps since 1973. This was not a full list and since this first meeting the British side has pressed the Libyans for further names. Libyan-trained terrorists are the core of the IRA and are known to have been responsible for a large number of killings since 1979, when men from this group murdered Lord Mountbatten.

That the IRA threatened to kill Gaddafi for betraying the organisation was perhaps an indication of the value of the information that he passed to British Intelligence.

The War in 1993-94

As always, the IRA continued to insist that it wants peace while at the same time planting bombs in public places. Simultaneously, the organisation conducted a propaganda campaign, aimed mostly at world opinion. In February 1993 the IRA contacted the British government, 'looking for advice', as its spokesman said, 'on ending its campaign of violence'.

On 22 March a deputy from Sinn Fein, the political wing of the IRA, secretly met a third party who told him that the Northern Ireland Secretary, now Sir Patrick Mayhew, had made an offer that would allow Sinn Fein to take part in general talks if the IRA would not only end its violence but openly renounce it.

On 10 April, John Hume, a moderate Irish nationalist and leader of the Social Democratic and Labour Party (SDLP), began secret talks with Sinn Fein's Gerry Adams. On 29 October John Major and the prime minister of Eire, Albert Reynolds, agreed to work together but they rejected the Hume-Adams proposals. In December, Major and Reynolds met in London to announce their plan. They signed a joint declaration in which Britain for the first time formally acknowledged the possibility of a united Ireland if a majority of people in the north agreed.

Dublin's offer to relinquish its claim to Northern Ireland under the Irish Constitution as part of any overall peace agreement was also formally enshrined in the document. The two leaders than placed the onus on the IRA, telling its leaders that exploratory talks could begin in three months if the violence ended.

Following the peace declaration the Pope made a personal appeal to the IRA to end its campaign of violence, but within a month the IRA was launching rocket attacks in Belfast.

Sir Patrick Mayhew brought up the controversial issue of the IRA's huge arsenal. The mechanics of how to hand over the arms could be discussed during exploratory talks, he said. However, he once again stressed the basic British position: 'We are not in the business of bargaining with the IRA.'

Mayhew also said that the British-Irish call to end violence applied equally to the Protestant extremist organisations, which were responsible for half the 95 violent deaths in Northern Ireland in 1993.

Sinn Fein's initial response to the Major-Reynolds initiative was to say that it was 'disappointed' but it promised to study it further. Hardline Protestants, who want the province to remain British, dismissed the initiative as a 'sell out'.

Even while the IRA was supposedly studying the peace proposals, its active service units continued their bombing and assassination campaigns. The reason was unclear. It is possible that Adams and IRA leaders are unable to control rogue units that are determined to continue the armed struggle against the British. Equally possible, the attacks were part of Adams's negotiating strategy. Adams has done nothing to explain the anomaly of his talking peace while the IRA continues to make war - 'bullets and ballots approach'.

In November 1993 British security forces intercepted a large shipment of arms and ammunition, including 300 assault rifles and two tons of explosives. Members of the Ulster Volunteer Force, an illegal Protestant paramilitary group, admitted that they had bought these arms in Eastern Europe and declared that they would continue to 'scour the world' for more. Peter Robinson, a hard-line Unionist politician and parliamentary member for the East Belfast district for which the arms were destined, said: 'People are preparing themselves for war.'

War was evident on many occasions. Below is a selection of the more 'important' events in 1993:

20 March: An IRA bomb in Warrington, Lancashire, kills two children and wounds 56 people. The murders arouse international revulsion.

24 April: An IRA bomb in London's financial district kills one, wounds 45 and causes £1.5 billion worth of damage.

23 October: An IRA bomb explodes prematurely beneath the West Belfast HQ of the Ulster Defence Association, killing nine members of the public, mostly in a fish shop, and one of the bombers.

25-30 October: Loyalist gunmen murder 13 people in Catholic areas.

In March 1994 the IRA carried out one of its most successful coups when active-service units targeted Heathrow Airport, London, with mortars concealed close to the runways. Nobody was injured since the bombs did not explode. However, it was a major success for the IRA because it indicated that it could operate in Britain with impunity.

The IRA might decide that more killing, together with secret 'quasi-talks' with the British, will win more concessions. Equally, it may find more killing difficult to justify with nearly all nationalists, north and south of the border, finally on the same side as the British government. The sad fact is that the men of violence need no justification; their murderous hatred is ingrained.

Papua New Guinea

ISLAND REBELS

The 160,000 inhabitants of Bougainville Island and the smaller Buka Island, off the eastern end of the great island of Papua New Guinea (PNG) have sought independence since 1989.

When the PNG government refused to consider the secessionist request, the Bougainville Revolutionary Army (BRA) came into being to bring about independence by force. In March 1990, the PNG authorities withdrew their troops following a bloody uprising but, when this encouraged the rebels to set up their own administration, the troops returned, though kept in fortified camps.

The government applied a tight blockade to the island in July 1990. No medicine or hospital supplies have been allowed in and the health of the islanders has deteriorated alarmingly. Diseases which had been virtually conquered have re-emerged and people are dying from malaria, yaws and beri beri. Mortality among the 10,000 young children is the highest in the Pacific.

In October 1991 about 40 armed rebels from the BRA's stronghold of Kieta, as well as from islands to the north, made an attack on the PNG security forces' camp in north Bougainville. The troops were part of a unit stationed on the island, supposedly to implement a government programme to restore services.

The attack surprised the garrison but, according to government source, only six of the soldiers were wounded while 15 rebels were killed. The PNG Defence Minister, Benais Sabumei, appealed to chiefs in northern Bougainville to continue to work with the troops in restoring services but the BRA regards the chiefs as collaborators.

Three rounds of peace talks have been held, in efforts to bring peace to Bougainville but the two sides have not been able to agree on a venue or time for the discussions. The government rejects a demand by the rebels that the talks be held in a neutral country, possibly New Zealand.

Bougainville, which is mountainous and jungle-clad, is highly suitable country for guerrilla activity. During the Second World War the Japanese army of occupation failed to capture Australian coastwatchers and the invaders suffered heavy casualties at the hands of Australian and native guerrillas. The PNG security forces are much smaller than those of the wartime Japanese and they have fewer aircraft and ships. The BRA could hold out for decades and continue to inflict losses on the PNG troops and on the government. The government's main weapon will remain the blockade, which is designed to put pressure on the civilian population. Their suffering, the government hopes, will bring the rebels to a point of surrender.

As with so many other larger disputes throughout the world, that in Bougainville is made more complex by internal differences. The people of south

Bougainville see their area as a southern State within an independent Bougainville. Like the people in the BRA stronghold of central Bougainville, the southerners are adamant that any agreement reached must include the total and permanent withdrawal of the PNG army, but they are less hardline in their approach to negotiations with the government. According to documents leaked from the PNG Defence Intelligence Branch, the government's strategy is to play on the differences of opinion between central Bougainvilleans and those in the north and south. According to these documents:

> The split in the province will grow and it can become explosive if services are completely cut off. The government must cut off shipping, deliberately to worsen the hardships the people are facing. Simultaneously, a psychological warfare effort must go into action to exploit the situation.

Australia's Role

The third party to the Bougainville tragedy is Australia, which gave $275 million in aid to PNG in 1991, $38 million going to the Defence Force, which is generally disliked. Australia is reviled in Bougainville for giving the PNG forces the four helicopters used in the so-called St. Valentine's Day massacre in February 1990. Australia said that it had given PNG the aircraft on the understanding they would not be used as gunships, but the Defence Force did so employ them, and many people were killed or wounded.

There have been numerous atrocities — well documented by Amnesty International — since fighting began in 1989 after local anger when the Bougainville copper mine, which was largely Australian-owned, could not continue to operate. The Hawke government was under much pressure to 'do the right thing', regardless of the diplomatic consequences. For many people the 'right thing' is to put sufficient pressure on the PNG government to force it to allow medical supplies and food through to the suffering natives of Bougainville.

Given Australia's great regional influence and the military aid it supplies to PNG, it obviously has a critical responsibility in the dispute. In fact, Prime Minister Hawke and Foreign Minister Senator Gareth Evans made personal representations to the PNG Prime Minister Rabbie Namaliu about the blockade. There is, however, a serious political problem: Australia could not intervene militarily or in any other manner which would lead to its being branded 'neo-colonialist'.

According to Hank Nelson, a senior Fellow at Australian National University and an authority on PNG, 'whatever Australia does, it has to support the central government'.

In November 1991, the Australian government was accused of engaging in political censorship in refusing to issue a Bougainville church leader with a visa. Bishop John Zale, of the United Church of Bougainville, was invited to Australia by churches and aid agencies to address meetings on the human cost of the PNG military blockade. The tour was cancelled when Bishop Zale was refused an entry visa to Australia.

Australian critics echoed complaints by the BRA that Australia was denying the rebel organisation the right to put its case for secession publicly. But the Minister for Foreign Affairs, Senator Evans, said that Australia could not issue the Bishop with a visa because he did not possess a valid PNG passport. Late in October, the PNG government cancelled the passport of Bishop Zale and other senior secessionist leaders.

Papua New Guinea's military strength

Papua New Guinea has a population of more than 2.5 million but total armed forces of about 3,500. The army, with a strength of 3,200, consists of two infantry 'regiments', though they are in fact only battalions, together with one engineer battalion. The navy has 300 personnel and the air force merely 150. To enforce its blockade of Bougainville, the navy has 4 *Tarangau* class patrol and coastal boats and two amphibious craft, *Salamaua* class. All were built in Australia. Australia maintains 110 military advisers and an engineer unit of 70 in PNG.

Critical Capture

In December 1993 the Papua New Guinea government took the interesting step of announcing the manoeuvres of the PNG Defence Force towards recapture of Panguna and the abandoned Panguna mine. The office of the Prime Minister broadcast its military intention. 'Security forces on Bougainville have moved close to the mine which the government wants to control by the end of the year.'

Soon after this, a spokesman for the Minister for Bougainville Affairs, Michael Ogio, said that BRA rebels were believed to be in the area waiting to ambush the PNG Defence Force. 'Our soldiers are not in an offensive approach', he said. 'They merely want to control the area.'

This 'war by radio' tactic at least had the effect of inducing civilians in the area to flee to government-run care centres. It did not frighten the BRA forces but in the subsequent fighting they were beaten and after four years of fighting the PNG Defence Force won back the mine and the area around it. This operation did not end the war and desultory fighting is likely to continue until Bougainville wins its independence.

Peru's 'Shining Path' War

A COUNTRY IN CHAOS

Background Summary

The name *Sendero Luminoso* (Shining Path) is part of the slogan of the Communist Party of Peru, which was founded in 1980 by Abimael Guzman, a professor of philosophy. Hardly anybody uses the term Communist Party, so *Sendero Luminoso*, from 'the shining path to Communism', is the accepted name.

In practice, Guzman had founded the party 17 years earlier. As a university lecturer, he was training schoolteachers and guerrillas from the early 1970s when Peru was under military dictatorship. He came to prominence in 1980 when he made a great show of destroying the ballot boxes in the village of Chusci, near Ayocucho. Through this symbolic act, Guzman declared war against the state and his subsequent actions showed that he was intent on creating a worker-peasant state based on Maoist ideals.

In the course of *Sendero Luminoso*'s violence and the security forces' counter-violence, about 15,000 people had died by the end of 1989. Alan Garcia was elected president in 1985 and faced appalling problems, not least the inefficiency of his generals and the brutality of his soldiers. In 1988, the Rodrigo Franco Command (CRF), a Right-wing death squad, made its appearance and set about killing academics and others who showed sympathy for rebel movements.

Sendero Luminoso was equally ruthless in 1989, killing 696 political candidates, public officials, soldiers and policemen and intimidating thousands more.

Summary of the War in 1990

President Garcia led Peru down a road of mismanagement and overspending that worsened its economic crisis. After two elections he was replaced by Alberto Fujimoro, the 51-year-old former Rector of Lima's Agrarian University. As he prepared to take office, Left-wing guerrillas of the Tupac Amaru Revolutionary Movement (MRTA) fired mortar bombs at the presidential palace and *Sendero Luminoso* blacked out most of Lima by blowing up electricity pylons.

Throughout 1990, *Sendero Luminoso*'s influence dominated most of the Peruvian Andes, many coastal valleys and large parts of the jungle. Desperate peasant groups formed village defence groups but the terrorists continued to win control whenever 'natural support' was lacking. Even so, the security forces used terror to an even greater extent and, according to the UN, in the period 1987–90 Peru led the world in the number of people to have 'disappeared' while under detention.

In May 1990 the US government announced a programme of military aid for Peru, including the establishment of a training base to be run by US Special Forces in the main area growing coca leaf, the source of cocaine. Under the plan, the Peruvian Defence Ministry received $35 million in 1990. The American ambassador in Peru, Anthony Quainton, proposed that the herbicide 'Spike' should be used to kill the peasants' coca crop. This misguided policy was to give *Sendero Luminoso* the boost its leaders hoped for by turning the great mass of peasants (who grow coca for their livelihood) towards *Sendero Luminoso* as their protector against the Lima–Washington policies.

The War in 1991–92

The movement intensified its attempts to force religious organisations to drop their humanitarian activities. In May, it held a 'people's trial' in the highland village of Huasahuasi in Junin department and tried five missionaries and social workers for 'crime against the peasants'. Then it shot them through the head. One of the victims was Sister Irene McCormack, an Australian.

In the same week a guerrilla column in Junin assassinated a provincial director of the Catholic church's national charity. A Canadian evangelical pastor, Norman Tattersall, director of the US-based aid organisation 'World Vision', was gunned down in front of his Lima headquarters. The killings during that week brought the total for the war to 22,000.

The churches have been called 'the state's front line' and whether or not this is true in any political sense, they are certainly perceived as such by *Sendero Luminoso*. This is why the movement is both attacking and infiltrating the church and other popular organisations. During 1991 the terrorists targeted the desert valley of San Juan de Lurigancho, where perhaps one million of Lima's population of seven million live. Police in the valley have retreated in the face of *Sendero Luminoso*'s advance. Only three police stations are manned and those only by day. Across Peru, hundreds of police stations are deserted and in many areas the police patrol only the main roads; other roads are too dangerous.

Sendero Luminoso's grip on the Lima districts is badly handicapping industry and business. A number of companies and hotels have closed down following death threats made against their managers. Many Lima factories erected electric fences and gun turrets during 1991 but the terrorists still manage to display posters extolling the virtues of 'President Gonzalo', the movement's name for Abimael Guzman.

Sendero Luminoso compels businessmen to pay up to $20,000 each for 'protection'. Otherwise, the guerrillas say, 'evil people will assassinate you and blow up your business'.

A foreign journalist asked a former minister of the interior, Agustin Mantilla, if *Sendero Luminoso* could one day bring down the government and perhaps replace it. 'If things go on as they are now, if we do not realise the danger we are facing, the answer is yes', he said.

The break-up of the state became evident in June 1991, when *Sendero Luminoso* 'celebrated' the 11th anniversary of the outbreak of their 'people's war'. They attacked a naval base, fired rocket-propelled grenades into the

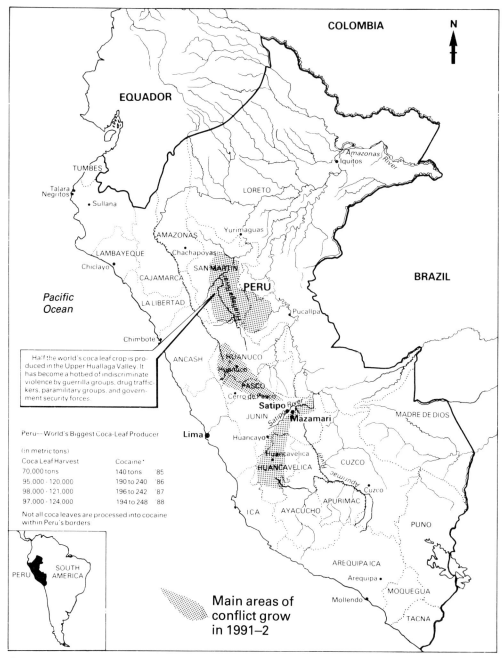

Peru: Sendero Luminosa (Shining Path) War

Finance Ministry, wounded six marines and killed a civilian with a car bomb attack, blew up a yacht and murdered a politician with a letter bomb. Scores of troops died in ambushes, peasants were massacred and guerrilla strikes paralysed many towns. Guerrillas themselves were killed but undoubtedly *Sendero Luminoso* had taken the initiative and had held it. Indeed, they claimed to have achieved 'strategic equilibrium' with the security forces.

Foreign military attachés admitted that the state was losing the war and this view was confirmed by retired Generals and Peruvian military analysts. President Fujimori was relying on intelligence to help him achieve some successes against the rebels but his anti-terrorist campaign was distinguished by nothing more significant than a drive to clean up *Sendero Luminoso* political graffiti in the universities.

According to Agustin Mantilla, the gravest problem posed by *Sendero Luminoso* was that its political strength was expanding all over the country and that its military strength could easily become commensurate. 'We are facing a ferocious war which is destroying our democratic system', Mantilla said. 'There is a loss of moral values, a loss of combativeness by the army and a retreat by the entire state apparatus.'

The rebels' dominance can be measured statistically. More than 60 per cent of Peru's population live in areas under a state of emergency which comprise 40 per cent of Peruvian territory. In the army, more than 1,000 officers resigned in a period of six months and 40 per cent of conscripts deserted. About 500 police were dismissed for corruption and for 'abandoning their posts' in the second half of 1991. The forces' morale is abysmal. The general elections failed to produce results in 30 per cent of Lima's districts because of a guerrilla boycott which intimidated voters into not going to the polls. When by-elections are called, no candidates offer themselves for fear of murder by *Sendero Luminoso*. Local officials resign *en bloc* when guerrillas threaten them with death.

The security forces, harassed and overstretched, also over-react and they torture and murder many people picked up as *Sendero Luminoso* suspects.

Sendero Luminoso has achieved a strategy stalemate, even if not the strategic parity they claim. Until the security forces are properly paid, better equipped and much better trained they cannot defeat *Sendero Luminoso*.[1]

A leaked document, in July, 1991, indicated that security forces in certain operations were under orders to take no prisoners. In early July, 24 people died in a massacre in Santa Barbara, an Indian village, about 155 miles south-east of Lima. Having finished their slaughter, the soldiers burned down houses and stole personal belongings and more than 700 animals. The massacre victims were said to be *Sendero Luminoso* sympathisers. The raid followed many others in Santa Barbara province and it indicated that the security forces were stepping up their dirty war. The leaked document is said to have originated in the Armed Forces Joint Command. It stated:

> In special intelligence operations, actions should normally be clandestine, using legal or illegal procedures. If the situation and the conditions permit, eliminations will be carried out without leaving any trace. Anti-subversive operations which are executed on the basis of information provided by the intelligence detachments will be of a highly aggressive and offensive

character, not forgetting that the best subversive is a dead subversive. Therefore no prisoners will be taken.

Sendero Luminoso command counts any over-reaction by the security forces as a victory, since it leads to a further breakdown in the administration of the state. Resentment, hatreds, fears and violence are what *Sendero Luminoso* feeds upon. Abimael Guzman and his lieutenants were doing no more than following the principles laid down by their understanding of Trotskyism and Marxism to bring down the machinery of the state. If *Sendero Luminoso*'s leaders had heard that the old ideologies had been discredited elsewhere in the world, they were taking no notice whatever.

The Great Capture

In September 1992 the government achieved what everybody in Peru had always believed was impossible - the security forces captured Abimael Guzman while he was staying at a house in Lima. With his arrest came predictions that *Sendero Luminoso* would collapse but, just in case Guzman might escape and reinspire the movement, the authorities imprisoned him on an island off the coast of Peru. Here he was incarcerated in a specially-built underground cell where the government was confident he would see out the rest of his life. The danger he represented was considered to be so great that even the members of the special unit responsible for escorting him to the island and the guards on the island were masked. These guards are masked at all times when in Guzman's sight. Occasionally he is exhibited to important visitors and to visiting cameramen as if he were a species of wild animal.

Sendero Luminoso's activities continued but with less intensity and in September 1993 President Fujimori announced that the organisation was 'in decline and on the road to complete annihilation'. Fujimori repeated his pledge to wipe out the organisation by 1995.

On 16 December 1993 *Sendero Luminoso* bombed a Lima bank and nine people were killed. The explosions shook the capital's sense of security. Worse was to come. On 28 December, three massive car bombs rocked Lima with earthquake-like intensity as *Sendero Luminoso* marked the 100th anniversary of the birth of Mao Tse-tung. One bomb levelled an air force funeral parlour while another wrecked a police complex and injured 28 people, most of them police. The third bomb destroyed the place where police had displayed Abimael Guzman after his arrest. A fourth bomb had been intended for the US embassy in Lima but police opened fire on the minibus racing for the embassy's gates. The terrorists driving the vehicle fled before detonating the bomb.

The casualties from the various bombs brought the total number of people killed in 13 years of civil war to 27,000. Total damage by the end of 1993 was estimated at $US 25 billion.

In a message which Guzman managed to smuggle from his cell - which in itself shows his enormous influence - he thanked his followers for their support and urged them on to 'greater exploits in the name of freedom'.

During 1994 they responded with innumerable attacks on the police, the army, village leaders considered hostile to *Sendero Luminoso*, wealthy Peruvians and many other targets. The momentum begun by Abimael Guzman was continuing to the Maoist plan.

References

1. 'People are desperate and you can actually see them getting thinner because of shortages', said a European diplomat. This desperation can be understood when, for instance, the wages paid to teachers and health workers average $35 a month. About 350,000 of them were on strike for some months during 1991.

War Annual 5 carried details of human rights abuses and an analysis of the 'misguided' American military aid to Peru.

Philippines 'People's War'

A NEST OF CONFLICTS

Background Summary

In 1969 the Communist Party of the Philippines (CPP) and its military wing, the National People's Army (NPA) began what they called a people's war against the corrupt President Ferdinand Marcos and his oppressive regime. The NPA's strength was never more than 20,000 but it was active in 60 of the 73 provinces and it had many minor successes. Samar Island, 200 miles south-east of Manila, the national capital, was the NPA's strategic centre.

Various private 'armies', all of them Muslim, also harassed the government's forces. The main groups were the Bagsa Moro Army, the fighting wing of the Moro National People's Liberation Front (MNLF), the Moro Islamic Liberation Front (MILF) and the Cordillera People's Liberation Army (CPLA). Casualties were heavy, though the security forces lost more men than the elusive guerrillas.

Marcos was replaced by President Cory Aquino, but the war continued. In 1989 the Defence Secretary, General Fidel Ramos, adopted a new strategy. Out went the old search-and-destroy methods and in came the seven-person Special Operations Teams (SOTs), which used political and psychological methods to dissuade the peasants from supporting the guerrillas. While the government had success with the SOT programme, and with conventional warfare in some areas, the MNLF, under Nur Misuari, began an offensive in support of its claim for a separate independent state for the Philippines' eight million Muslims.

However, the government's main military problems were caused by Lieutenant Colonel Gregorio ('Gringo') Honason, a clever renegade, who attracted other officers and soldiers to his anti-government campaigns. He founded the 'Reform the Armed Forces' movement (RAM) which attracted many officers.

Colonel Honason's operations against the government started in earnest in September 1990 when RAM began an uprising in Mindanao, 'the first step of a great historical process towards realising a federal-parliamentary form of government for the entire Philippines archipelago'.

Mrs. Aquino resisted pressure to declare martial law, probably aware that for many Filipinos martial law is synonymous with the autocratic rule of President Marcos and for that reason hated. Some senior officers, however, contend that the armed forces are more representative of the people than Congress itself.

The War in 1991–92

Disputes over the continued existence of the two great US bases in the Philippines — the navy's Subic Bay and the air force's Clark Base — remained a basic source of conflict. The removal of these large bases, together with four smaller ones, had been demanded for many years. The NPA naturally wants the bases removed because they constitute emergency aid for any government facing a threat from insurgents. During 1989–90 Mrs. Aquino called on the US for military aid in quelling army mutinies. The NPA offered to declare a truce in its 'people's war' if President Aquino declared that the bases would be closed down in September 1992 when, in fact, the US–Philippines Treaty governing them runs out. The NPA also said that it would no longer kill American servicemen.

On 11 September 1991, the National Democratic Front, which includes the NPA, did declare a ceasefire when a treaty extending the lease of the Subic Bay base faced defeat by the Philippines Senate. Soon after, the majority of the 23-seat Senate voted against the treaty. However, the leaders of the Philippines armed forces did not respond to the ceasefire offer, saying that the Communists always violated their own ceasefires and that they were therefore meaningless.

President Aquino then said that she might call for a referendum on extending the bases' leases. In addition, she provisionally extended the leases by withdrawing an order requiring the Americans to pull out in September 1992. The President was influenced by economic and political considerations. The bases employ 70,000 Filipinos fulltime and are worth several billion dollars annually to the Philippines economy. The central Luzon sector, where Subic Bay base is located, was devastated by the eruption of Mount Pinatubo in June and July and it needed the support which American money provided.

In addition, the US State Department and the Pentagon were putting pressure on pro-US Philippines senators to back the continuation of the American bases. With instability in South-East Asia, the Indian sub-continent and parts of Indonesia, the American military planners believe that they could need the great bases in the Philippines until the mid-1990s. Senate Foreign Relations Committee chairwoman Ms. Leticia Shahini said in September 1991 that the US might be allowed a withdrawal period of seven years.[1]

The National Democratic Front and its constituent groups announced that their ceasefire was now null and void and that operations against the government would continue. The first casualties were inflicted on government forces soon after midnight on 22 September.

A week later Colonel Honason appeared before a clandestine news conference to say that his RAM movement was offering peace talks in return for political change, including immediate withdrawal of the American forces. The alternative was further bloodshed, he said: 'if this government proves once again, as it has done in the past, that it understands only one language, then we reserve the right to use the same option. We have demonstrated that we have significant military capability'.

Colonel Honason said that RAM had dropped an unrealistic and earlier demand for the resignation of President Aquino. He rejected government suggestions that the rebels were a spent force and warned that RAM was now

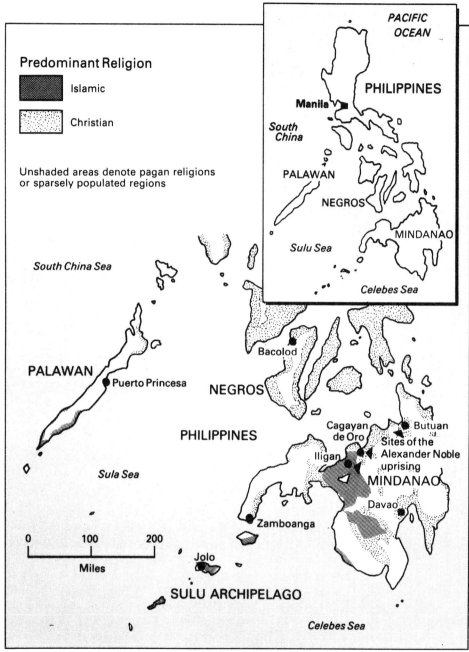

Philippines Ethnic Areas

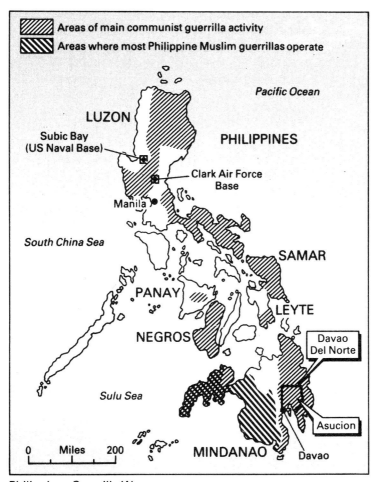

Philippines Guerrilla War

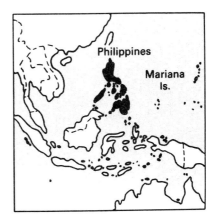

stronger and could launch a fresh 'putsch' at any time.

He wanted government action on eight 'talking points'. They included electoral reforms to ensure a free and fair ballot in national polls scheduled for May 1992, eradication of corruption in government and the immediate return of the late President Marcos from Hawaii. In fact, Honason was one of the first people to greet Imelda Marcos, widow of the former president, when she returned to the Philippines in November 1991.

However, just before this significant event a leading army rebel, former Lieutenant-Colonel Eduardo 'Red' Kapunan, appeared in Manila, surrendered to the government and promised not to participate in any future coup attempts. Thirteen other officers, some of them senior to Kapunan but less notorious, also surrendered. President Aquino invited Kapunan for breakfast in the presidential palace, a move calculated to show that she bears no ill will towards rebel officers who surrender. The conciliatory gesture is unlikely to attract many more of the hundreds of officers who have joined rebel movements. On 1 July 1992 the former defence minister, Fidel Ramos became President of the Philippines in succession to Mrs. Aquino. He is expected to pursue an even harder line with the rebels than Mrs. Aquino did.

In the meantime, fighting takes place in northern and southern Luzon, in all of Negros and much of Samar and eastern Mindanao. In far western Mindanao, Communist guerrillas and Muslim guerrillas operate, sometimes against each other. The government works hard at playing down the scale of the conflict. During 1989–90, the government and armed forces press offices were eager to help correspondents to tour the various islands and the campaign areas to see for themselves the success of the anti-guerrilla operations, whether by SOT teams or by aggressive large-scale patrols. However, in 1991–92, as the army became demoralised and as the influence of Honason's RAM spread throughout the units, the officials became less co-operative, several foreign journalists have described them as obstructive.

Defecting soldiers are supplying the NPA with mortars and rocket-launchers as well as ammunition. It is these supplies which give Honason and his colleagues confidence that they can so seriously damage the security forces that the government could fall. RAM is trying to buy artillery pieces and, according to unconfirmed reports, at the end of 1991 some guns had been shipped into mid-Luzon from North Korea.

References

1. Ms. Shahini said that before a referendum could be called, it was necessary, under the Constitution, to gather more than three million signatures in order to present a petition for a referendum. It should be noted that the main instigator of the moves to extend the lease is Emmanuel Pelaez, the Philippines ambassador to the US.

War Annual 5 carried details of Colonel Honason's tactical operations against the government and a lengthy account of the American bases dispute.

South Africa: Zulu — ANC War

THE 'BLACK-ON-BLACK' CONFLICT

Background

The Zulus of north-east Natal and the Xhoso-speaking people in the east of Cape Province have been warring against each other for more than 200 years. In modern times, political party organisations have been contemporary fronts for old tribal rivalries. The conflict between the Zulu Inkatha movement and the African National Congress (ANC) since 1976 merely reflects those animosities. Chief Mangosutha Buthelezi, who founded Inkatha in 1976, is Chief Minister of KwaZulu, the tribal 'homeland' of South Africa's seven millions Zulus. He has quite different policies from those of Nelson Mandela, a Xhosa, who leads the ANC. For many years Zulus have supported Inkatha almost to a man while Xhosa people backed the ANC.

Nearly all ANC members regard Buthelezi, who formed Inkatha after having worked with the ANC, as having sold out to the whites. They accuse him of having abetted apartheid by serving as Chief Minister of KwaZulu, one of 10 'homelands' created by the South African government, where blacks exercise limited political rights. The ANC has always condemned Buthelezi for opposing both 'armed struggle' and international sanctions against Pretoria.

As leader of the largest tribe in the country and in control of the trained and motivated Inkatha army, Buthelezi is a key figure in South African politics. ANC, for its part, gained status with Nelson Mandela's release from prison in February 1990 and the progress he made in negotiations with South Africa's President, F.W. de Klerk.

The ANC recognised that it had to break Zulu power in Natal if it was to control all South Africa through a central government. One scheme began in 1986 when ANC agents infiltrated Zululand to recruit Zulu members. Since the release of Mandela, Zulus have lived in fear of being ruled by a member of the Xhosa tribe. At the same time, the ANC is intent on destroying Inkatha politically. Because Inkatha officially opposed violence against the whites — believing that co-operation would yield better results — many militant Zulus left the movement to join 'the struggle' under the ANC banner. Inkatha leaders tried to check this outflow by exploiting Inkatha's position as the homeland's administrator. They threatened reprisals against people who were not members of Inkatha, including eviction from their homes.

Following Mandela's release from prison, defections from Inkatha multiplied because many men saw the ANC as much more attractive. Inkatha responded in fury and the fighting intensified. The open and ferocious violence has few equals in any contemporary war. During 1990 the worst battlegrounds were the large migrant workers' hostels and the great squatter settlements. The hostels accommodate men who come from the homelands on short-term work contracts, leaving their families behind. The squatters, many of whom are unemployed, regard the temporary migrants as usurpers of jobs.

The bloodshed in August 1990 was the most concentrated upsurge of violence seen in Africa in the 20th century and the atrocious violence continued into December.

The War in 1991

A number of former South African Defence Force (SADF) officers said during 1991 that violence in the black townships was being orchestrated by military intelligence, with Special Forces doing the killing. These officers claimed that One Commando of Five Reconnaissance Regiment, generally known as 'Five Recce', was the most likely unit to be carrying out the numerous attacks in the townships. The professional nature of the attacks first persuaded the ANC to denounce the 'third force' which it considered was behind the killings. Five Recce was notorious in the 1980s for its counter-insurgency work and cross-border raids against ANC bases in neighbouring countries.

Strong evidence about Five Recce's complicity came from Felix Isias Ndimene, a former South African army sergeant, who served in the unit from 1983 until 1991. His revelations indicate convincing knowledge of the workings of Five Recce. He gave the names of a colonel, a commandant and a sergeant-major who presented the weekly 'political briefings', and the names of six soldiers who took part in a massacre, on 13 September 1990, on a Soweto-bound train.[1]

Ndimene confirmed what has long been known in military analysts in South Africa — and to some abroad — that Five Recce is a multinational regiment and, in effect, a mercenary one. Its members — 360 in mid-1991 — were from Mozambique, Namibia, Zambia, Zaire, Zimbabwe, Eire and Australia. Spokesmen from SADF sought to discredit Ndimene's testimony but I believe it to be true. He said in July 1991 that about 120 Five Recce soldiers had been involved in operations in 1990–92 in the Johannesburg area and in Natal Province.

The scale of the war can be seen from what happened in Indaleni, a large town in Zulu-dominated Natal Province. In March, a chieftain allied to Inkatha sent his warriors to 'purify' the town; this meant that he wanted ANC dominance eradicated. The warriors drove out all 40,000 inhabitants, killing or wounding any men who resisted.

By watching the abandoned town, ANC fighters — known among themselves as 'comrades' — learnt that the chief's Zulus came at night to loot the homes and buildings. One night in April, the 'comrades', armed with firearms, set up an ambush and in the ensuing fighting 50 men died, most of them Inkatha. At this point the South African police intervened, arrested 24 ANC men and charged all of them with murder. Observers thought it strange that not one Inkatha man in the area had been arrested, although all the violence, until the night of the ambush, had been created by Inkatha. Indeed, Inkatha warriors were taking over or 'purifying' one ANC settlement after another. The strong inference was that the police were aiding and abetting Inkatha.

Early in May the government announced that a 'peace summit' would take place in an attempt to end the Inkatha–ANC war. It was duly held on 24 May and government ministers and Chief Buthelezi's representatives turned up. The ANC failed to attend, on the ground that the conference was loaded against it. Instead, the ANC church leaders set up an alternative peace summit at which the ANC spokesmen were the most prominent speakers, though the government and Inkatha also sent delegates. Nothing was achieved at either meeting and in the third week of June any peace that might have existed was destroyed by a renewed outbreak of violence. In the space of 48 hours three massacres took place. In one, a family of six was butchered and in another 12 people were shot dead. On a morning train carrying workers bound for Johannesburg, a gang with firearms and machetes killed six people and wounded another 20. It was the 24th train massacre in 10 months and the finger of suspicion pointed at Inkatha in all cases.[2]

Investigations by the BBC and *The Independent* newspaper, London, revealed clear evidence that Inkatha and the security forces fomented killings in Johannesburg and in Natal.[3] The inquiries showed that the police not only turned a blind eye to Inkatha atrocities in Natal but on occasions actually took part in the murders. A man called Siphio Madlala confessed that he had worked for military intelligence in Natal, that he had been seconded to the security police and that he had taken part in hit-squad killings of senior ANC officials. The BBC–*Independent* investigation found 'overwhelming evidence' of instances where police in armoured cars escorted the Zulus to their killing fields and back to their hostel bases.

More evidence emerged of links between the South African state and the war in Natal when Jacques Van der Merwe, a former member of the Civil Co-operation Bureau (CCB), spoke out as a matter of conscience. The CCB was actually a SADF death-squad which was closed down after it had been exposed. Van der Merwe admitted to taking part in the killings of four people in Namibia during 1989. He reported that an army colonel had several times offered him 'contract work' in Natal — that is, killing of ANC officials.[4]

In August 1991 the National Peace Initiative (NPI), which was backed by most South African churches, finally induced the government, ANC and Inkatha to accept a peace accord. It includes codes of conduct for the political parties and the security forces, and an investigation into the causes of the violence. The accord was the latest step in a process begun in June and nurtured by church leaders, notably Archbishop Desmond Tutu.

On the night of 26 July 1991 several Casspirs (big police personnel carriers) arrived in the small black township of Kwadela in eastern Transvaal. Armed men, both police and Inkatha, dismounted to attack the homes of local ANC branch committee members. After nearly 30 people — men, women and children — had been killed or wounded, the Casspirs drove around the town to collect the killers and take them back to their bases. The Kwadela affair, though by no means isolated, became well known because of the evidence collected by human rights groups.[5]

As late as December 1991, South Africa was thrown into another political

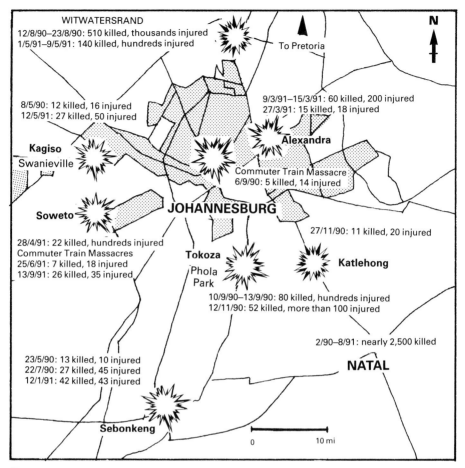

Zulu – ANC Battlefields

scandal after the *Weekly Mail* published (13 December) fresh allegations of extensive government funding of Inkatha death-squads. The newspaper stated that the SADF, through its military intelligence division, had channeled more than $3 million since 1987 into training a 200-man hit-squad. The worst atrocity in 18 months occurred on the night of 17 July 1992 when Zulus of Inkatha rampaged through the township of Boipatong. They massacred 39 men, women and children and wounded scores of others. According to all of the 30 reporters who visited Boipatong, police vehicles delivered some of the killers to the township. The atrocity caused great unrest and forced the ANC to withdraw from the peace process.

Despite the continuing war in South Africa, there was an optimistic scenario at the end of 1992. These were its phases as proposed by its advocates:

- An all-party conference would negotiate towards a new constitution. The problem was that the Conservative Party, which holds 22 per cent of seats in the White chamber of parliament and clings to White supremacy, might boycott the conference. The Afrikaner Resistance movement would try to sabotage the negotiations.
- Two major power blocs would emerge. The first would include the ruling National Party, which held 5 per cent of seats in the South African Parliament and had significant Coloured support: Zach de Beer's Democratic Party, 18 per cent of seats; and Inkatha, which had a lot of White backing. The second alliance would consist of the ANC; Clarence Makewetu's Pan-Africanist Congress; and Pandelani Nefolvhodwe's Azanian People's Organisation.
- A new constitution would be drafted, beginning in 1992. The ANC demanded and the government agreed that elections be held for a constitutional assembly so that voters could directly choose those who would write the constitution.
- A new, fully democratic South Africa would then be born as a result of elections to be held for an integrated multi-racial parliament.

Conflict During 1993-94

The process of negotiating peace and democratic elections began formally on 20 December 1991. There followed a hiatus but a Record of Understanding was signed in September 1992. Again there was a hiatus but the process resumed on 1 April 1993 and finally elections were scheduled for 26-28 April 1994. It was hoped that a government of national unity would then follow. One political analyst, Frederik van Zyl Slabbert, compared the ANC and the National Party as two drunks who had to prop each other up. He meant that after the elections the two sides would have to resume the often reluctant partnership that had enabled them to escape from the terrible burden of the apartheid past.

The pre-election period of 1992-94 was bedevilled by fighting between the young 'comrades' of the ANC and the Zulus of Inkatha and by equally bloody fighting *within* the ranks of Inkatha and of the ANC. For instance, in January 1994, infighting between rival gangs of ANC supporters plunged the town of

Katlehong, 20 miles south-east of Johannesburg, into new depths of savagery. The battle was between the township's self-defence units, created by the ANC to provide protection from Inkatha, and the South African police. In the depths of despair, Nelson Mandela accused the President, F.W. De Klerk, of allowing the violence to continue as a 'deliberate strategy' to undermine the ANC's electoral chances, an allegation of little substance.

The new government after the elections is likely to be threatened by the dormant black anger in the vast shanty towns, if only because Mandela and his new government cannot possibly meet their followers' high expectations. The blacks want land and property. They are unlikely to get as much as they want, as quickly as they want.

However two other groups threaten any vision of South Africa as a peaceful democracy. One is the Afrikaner Right wing led by General Constand Wiljoen; the other is the Inkatha Freedom Party of Chief Buthelezi. King Goodwill Zwelithini, Buthelezi's nephew, comes into the picture too. He wants the Zulu kingdom restored to its 1834 boundaries. For the million Zulus, land and hereditary rule are inextricably mixed and neither Buthelezi nor Zwelithini can easily be bought off with promises of a constitutional monarchy in a federation in which central control rests with the ANC.

Throughout 1993 and the first half of 1994 the hate between the ANC and Inkatha was grotesquely apparent. Various efforts to bring about peace between them foundered in violence. For instance, early in April 1994 a group of nine ANC men visited Kwamashu, a town near Durban, on a goodwill mission. To talk peace, they visited an Inkatha hostel. The Zulus bundled them into a van, took them to a nearby railway station and shot them one by one. Only one survived. The slaughter was in revenge for the killing of Zulus the previous day when Inkatha warriors and ANC security guards fought a running gun battle through Johannesburg's commercial area. Chief Buthelezi could hardly have been more inflammatory. He announced: 'This is the final struggle between the ANC and the Zulu nation.'

President De Klerk responded by declaring a state of emergency in Natal Province, which includes the KwaZulu homeland where Buthelezi is chief minister, but warfare raged in the final weeks before the elections. Early on 4 April thousands of Zulus carrying spears, axes, club, pistols and some AK-47s stormed into Johannesburg's business district. At Shell House, the 21-storey office building housing ANC headquarters, security men fired a fusillade into the demonstrators, creating a jumble of bleeding bodies. A few blocks away, rooftop snipers opened fire. The casualty toll for the incident was 53 dead and 450 wounded.

The long conflict between Inkatha and the ANC has many roots but at its core is a power struggle between the Congress, which Buthelezi sees as Marxist and dangerously revolutionary, and Inkatha, which the ANC depicts as a Right-wing ethnic party led by an autocrat. Buthelezi, the autocrat, was the only significant political figure in South Africa, Right or Left, who refused - until the last week - to take part in the elections. He said that he was holding out for a federal system that would keep the ANC from dominating the Zulus. If he could not win provincial autonomy he demanded a sovereign Zulu kingdom under the rule of King Goodwill. The future of South Africa, with so many disparate forces with radically different agendas, can only be turbulent.

References

1. Ndimene spoke to the *New Nation* newspaper. This report was picked up and reported by veteran correspondents in South Africa, most of whom believe Ndimene's allegations. He said of the 13 September train massacre: 'My friends got on the train with pangas and AK-47s and they were using the name of Inkatha. They shot the people and killed them with the AK-47s. They said they were not allowed to speak during that attack because most of the people were Namibian and could not speak Zulu.' Many reports from the townships say that black men speaking non-South African languages and with non-South African features, have been seen at the scene of killings.
2. A study of the Community Agency for Social Inquiry, an independent academic body, found that Inkatha had been the aggressors in 66 per cent of the acts of violence in Johannesburg townships, the police in 13 per cent and the ANC in six per cent.
3. *File on Four*, BBC Radio 4, 23 July 1991, presented by John Carlin.
4. The Johannesburg *Weekly Mail*, 31 July 1991, showed that the government, despite repeated denials, had provided covert funds via the police to support Inkatha. Pretoria admitted that Inkatha and an allied labour union received more than $600,000.
5. Max Coleman, a commissioner of the Human Rights Commission, based in Johannesburg, said: 'What we are witnessing is a deliberate policy to destabilise the opposition — the ANC and all the liberation movements in general. And using whatever means are possible.' *Time Magazine*, 5 August 1991.

Sri Lanka Civil War

'TIGER, TIGER, BURNING BRIGHT'
Background Summary

This vicious civil war began in 1983 when the Hindu Tamils turned to violence in their attempts to obtain a separate state in northern Sri Lanka, to be called Eelam. The Buddhist Sinhalese, the majority population, opposed the idea of a separate state. The Tamils created several guerrilla groups, the best known, in 1983, being: Liberation Tigers of Tamil Eelam (LTTE); the People's Liberation Organisation of Tamil Eelam (PLOTE); the Tamil Eelam Liberation Organisation (TELO); the Eelam People's Revolutionary Liberation Front (EPRLF); and the Eelam Revolutionary Organisation and Supporters (EROS). In addition, there was a political, non-guerrilla group, the Tamil United Liberation Front (TULF).

The Tamils ambushed and wiped out an army patrol and, in retaliation, enraged Sinhalese mobs in Colombo murdered at least 1,000 Tamils. Many massacres followed, including that, in April 1986, of all TELO leaders by the Tigers. By this internecine atrocity, the Tigers ensured that they were masters of the resistance movement.

Heavily armed, through the help of the Indian Tamils and other Tamils worldwide, the Tigers developed a quasi-army structure so that when the Sri Lankan forces attacked their bases on Jaffna peninsula, in February 1987, the Tigers survived.

The President, J.R. Jayawardene, asked Prime Minister Rajiv Gandhi of India for help in rescuing Sri Lanka from chaos. Gandhi sent an Indian Peace-keeping Force (IPKF) but its presence was resented by many Sinhalese, notably the Maoist terrorist group, *Janata Vimukti Paramuna* or People's Liberation Front (JVP).

The IPKF's *Operation Pawan* cost it 2,500 casualties. The Indian troops were heavy-handed and Sri Lanka's new Prime Minister, Ranasinghe Premadasa, was anxious for the Indians to go home. The IPKF armed, sheltered and deployed members of the EPRLF, by now the only significant opposition to the Tigers. This resulted in yet more internecine fighting.

The JVP, operating mainly in the south, fought the government and from its stronghold in Akuressa it became so violent and insidious that the entire Sinhalese army of 33,000 was deployed against it. In retaliation against numerous JVP atrocities, Right-wing vigilantes formed death squads and in one month they murdered 1,000 JVP supporters. When the security forces captured and killed the JVP leader, Rohane Wijeweera, and five other leaders violence died down in the south.

By the end of 1989 the LTTE, under Vilupillai Probhakaran, was the dominant rebel organisation, having absorbed, taken over or crushed all the other Tamil organisations.

The IPKF venture, having totally failed in its mission, was called off. With 6,000 civilians, 800 Tigers and 1,200 of its own troops killed, IPKF began its withdrawal in September 1989. (The reasons for IPKF's failure are analysed in depth in *War Annual No 5*).

In 1990 the siege of Jaffna Fort, which was held by Sri Lankan government troops, was the major military event. The city of Jaffna itself was the Tamil capital and the Tamils were affronted by the Fort's being in the possession of a government garrison. After a siege lasting more than 100 days, the troops were brought to safety by a government relief column on 26 September 1990. In the West, we can see that the relief of the garrison and the controlled withdrawal was a military triumph but in Sri Lanka the siege of Jaffna Fort was generally regarded as a boost for the Tigers.

Massacres and retaliation massacres continued in several parts of the country. Journalists and lawyers were murdered and Amnesty International reported numerous atrocities and abuses of human rights. Having fought the security forces for seven years, the Tamils declared an indefinite ceasefire from midnight on 31 December 1990.

The War in 1991–92

Simultaneously with the Tamils' unilateral declaration of a ceasefire, President Premadasa vowed to restore peace. 'I promise to rebuild a land destroyed by violence', he said. 'Terrorism is being defeated; our vision is unity, not separation.'

The ceasefire lasted two days, which was a longer period than most seasoned observers would have predicted. On 2 January 1991, LTTE rebels attacked two army posts in the north-east, probably to show the government that it should not assume that it had defeated the Tamil separatists. This became all too apparent on 2 March when the State Minister for Defence, Ranjan Wijeratne, was killed by a car bomb. Wijeratne, aged 59, was the most powerful figure in the Sri Lankan government after Premadasa. Apart from holding the defence portfolio he was also Plantation Minister and general secretary of the ruling United National Party.

He had the backing of the armed services in his determination to crush the Tamil separatists and he threw his weight behind them in rejecting the Tamil Tigers' ceasefire at the beginning of the year. It had been Wijeratne who had induced the government to resume hostilities against the Tamils.

But the Tigers were not his only enemies. Wijeratne had directed all operations against the southern JVP. His crushing of their revolt left 20,000 people dead or missing. A week before his death he had begun a campaign against other 'enemies of the state' — the rapidly-expanding casino and gambling industry, largely controlled by Chinese from Singapore and Hong Kong.

Wijeratne was the third successive general secretary of the ruling United

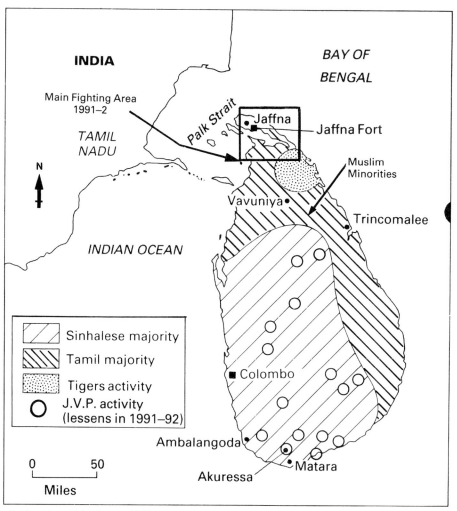

Sri Lanka: Ethnic and War Areas

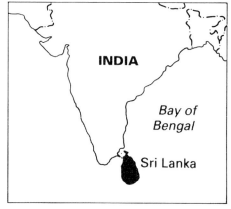

National Party to be assassinated since 1987. His two predecessors were shot by JVP killers. Wijeratne was passing through Havelock Town, Colombo, in his readily-identifiable white Mercedes, when the car bomb exploded, killing at least 19 other people and injuring more than 70. Tamil political groups who oppose the Tigers say that separatist guerrillas had enough explosives hidden in Colombo for a car bomb operation.

President Premadasa appealed for calm and warned that the security forces would take 'firm action' against anybody seeking to stir up trouble after the assassination; as a result, there were no demonstrations. Wijeratne, who held the rank of Colonel in the volunteer army, was posthumously promoted to General.[1]

In the last week of April 1991 the security forces launched an offensive against rebels on Karaitivu and Kayts islands and against other positions in northern Jaffna peninsula. In four days of fighting, the armed forces killed 126 rebels and wounded about 150, while suffering 32 servicemen killed and 70 wounded.

The Tamils counter-attacked almost immediately. About 60 soldiers from a camp in Nanaddan, north-west Mannar district, were ambushed when on patrol. At least 35 of them were killed in the trap and the army conceded that the attackers escaped unscathed.

Atrocities committed by one side were matched by the other. On 12 June three soldiers were killed in a landmine explosion which was blamed on Tamil rebels. Within hours, troops raided the Tamil villages of Makiladitivu and Manaikkadu, near Batticaloa. They set fire to homes and shot the occupants as they tried to escape. In all, 166 men, women and children died in the massacre; another 20 were wounded and 300 homes were set on fire.[2] The Joint Operations Command in Colombo flew doctors to the villages but it is unlikely that their report will be made public.

A senior Defence Ministry official, Air Chief Marshal Walter Fernando, trying to explain why the civilians were killed, said that they were caught in the crossfire when Tamil rebels tried to use them as human shields during a rebel attack on government troops near Batticaloa. 'If there has indeed been a massacre, the strongest action will be taken against the miscreants', Fernando said.

Fernando, promoted to the post of acting Defence Secretary, introduced *Operation Thunderbolt*, a security operation that involved the random check of many vehicles in Colombo for explosives and weapons. The operation failed and on 21 June a suicide bomber from the LTTE planted a car bomb close to the main military centre in Colombo. When it exploded more than 70 people were killed and about 200 injured. Fifty vehicles were destroyed and buildings in the military base were damaged. Great damage was also done to the army's prestige.

The Sri Lankan parliament extended the state of emergency which had been imposed in 1989. Anura Bandaranaike of the Freedom Party told parliament that the Tigers were highly motivated and well-armed. 'Make no mistake about it, you are not taking on school-kids', he said. The only surprising aspect of this utterance was that it was made at all. The Tigers had been showing their motivation and their pitiless ferocity since 1983.

The Battle of Elephant Pass[3]

At the beginning of 1991, about 500 soldiers held the head of Elephant Pass, a narrow 2km stretch of dunes and marshlands connecting Sri Lankan mainland to the Jaffna Peninsula, a Tiger stronghold.

Early in the year the Tamil commanders had decided to make a conventional assault on the entrenched army positions and, in well concealed preparations, they brought up guns and mortars and dug approach trenches in order to place their assaulting troops in position unseen. They tunnelled forward until they were close to the army compound, which was encircled by barbed wire.

Under orders from engineer officers, the Tamil soldiers built bunkers of railway sleepers and sandbags to protect themselves against the army's anticipated artillery fire, and fake outposts complete with uniformed dummies were strategically placed. The Tamil workshops turned tractors and bulldozers into armoured cars by covering them with steel plates. Towards the end of their preparations the Tamils brought forward their own rocket system, made in Jaffna, that could throw 50kg missiles a distance of 1,000 yards. Perhaps as many as 60 anti-aircraft guns were so placed as to prevent army helicopters from hovering over the camp to drop supplies and reinforcements and to evacuate the wounded.

The Tamils' security held until May when army intelligence reported that an attack was being mounted. As a result, the garrison's strength was doubled to 1,000 men.

On 10 July the assault began. About 3,000 Tamil fighters, including 500 women, moved rapidly forward along the trenches. Under cover of their own mortar fire and firing their AK-47 assault rifles and rocket-propelled grenades, they achieved tactical surprise. However, the outnumbered defenders ably fought back. Unable to evacuate his wounded, the base commander appointed a sergeant-major as surgeon and this man carried out amputations by following radioed instructions. Just why no surgeon was posted to a base which expected to be attacked has not been explained.

The Tigers experienced some setbacks. Their armoured bulldozers were too slow or they became bogged down in the sand. The Sri Lankan helicopter gunships, though unable to help the besieged garrison directly, prevented the Tamils from moving ammunition and food to the front.

After some days the determined Tamils were slowly but clearly winning the battle. Aware that a victory for the Tamils in their first excursion into open warfare would have disastrous effects on the morale of the army, the government and High Command made a desperate but none the less daring decision. Naval units landed 8,000 fresh troops and marines, under Brigadier Vijaya Wimalaratne, on a beach six miles from the beleaguered base. This relief column came under fire at once and could sometimes advance only 300 yards a day. Tamil fighters resorted to brave and suicidal headlong charges in their attempts to beat back the reinforcements. 'They came at us in waves', Brigadier Wimalaratne said. 'It was an amazing spectacle.'

After 24 days, the relief column reached the camp and broke the siege. Having suffered 564 fighters killed, the Tigers' leader, Vilupillai Prabhakaran, ordered a withdrawal. About 200 soldiers were also killed but the army, with

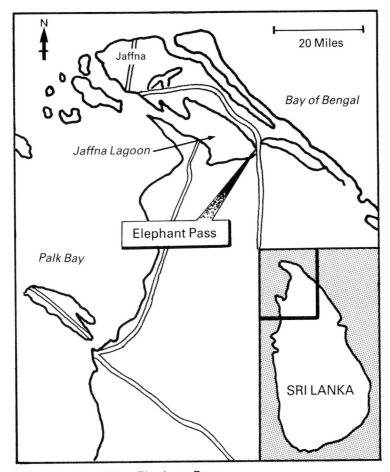

Sri Lanka Civil War: Elephant Pass

naval and air support, had won its first large-scale battle against the Tigers. Prabhakaran admitted that he and other commanders had failed to anticipate the amphibious landing but he claimed that the Tigers had won a moral victory. 'We have shown the world that we have evolved from a guerrilla force into one than can fight a conventional war against a modern army', he said. 'We had already learned much about the logistics problems of conventional war and now we know even more. We will fight our future battles better.'

As always, the guerrillas appeared to be unaffected by the reverse, even by the loss of several key commanders early in the Elephant Pass battle. The resilience of the Tigers yet again astonished observers. Even badly wounded men and women told foreign reporters of their eagerness and resolve to go back into the field to fight for a Tamil homeland. Some of this show of spirit may have been intended to deter the Sri Lankan High Command from exploiting its victory. If so, it failed in its objective. On 10 September the army launched *Operation Lightning Strike*, in an attempt to capture the Tigers' largest stronghold, deep in the Mullaittavu jungle of northern Sri Lanka. The assault did not at first succeed but according to the army more than 200 Tigers, including three 'area leaders', were killed in the fighting. In October the security forces managed to demolish one of the Tigers' main weapons factories in Mullaittavu. At the same time, Tamil units became trapped in a Sri Lankan minefield and suffered casualties.

Following these reverses, the LTTE asked — or ordered — all Tamils who had taken refuge or migrated to Australia, Canada, Britain, France, Switzerland and other countries to contribute to a fund to buy arms on the international market.

Major General Denzil Kobbekaduwa, in command of field operations against the Tigers, said: 'Nothing is going to stop us now. Our mission is to seek out the Tigers, kill as many as possible and destroy their fighting capability.'

The government campaigns have imposed hardships upon the Tamil communities. Government checkpoints and patrols stop supplies from getting through and almost everything is in short supply or rationed. Police and soldiers in government-controlled areas round up Tamils by the thousand. Young Tamil men are the main target of such sweeps and many have been killed, a fact well known to Amnesty International.[4]

General Kobbekaduwa's public show of confidence is understandable, as is that of his superior, Lieutenant-General Hamilton Wanasinghe, Chief-of-Staff of the Sri Lankan army. He said: 'We are prepared to conduct high-risk operations and take on the Tigers in their fortified positions.'

Operation Lightning Strike indicates this intention but its limited success shows the army's difficulties. Artillery fire and air cover are less effective in the jungle country where the guerrillas live and train and mostly operate. The Indian army could not defeat the Tamils and it went home virtually disgraced, yet it was stronger than the Sri Lankan forces. While the Tigers of Sri Lanka continue to draw financial, moral and material support from Tamil Nadu — the Tamil state in southern Indian — there is no chance whatever that the Sri Lankan forces can defeat the Tigers.

India is still helping Sri Lanka in its war against the Tamils. In September 1991 its forces stepped up pressure against them, especially along Palk Straits, a

20-mile stretch of water which divides Sri Lanka from southern India. The Indian High Command ordered its navy to shoot at boats, dinghies and any vessel suspected of carrying militants from southern Indian to north-east Sri Lanka. Sri Lankan naval ships were working with the Indian navy to isolate the Tigers. However, Tamil boats still manage to breach the blockade.

Turning Tamils into Tigers

The Tigers' 'secret weapon' is their discipline, as Vilupillai Prabhakaran himself has made clear: 'Commitment comes from strictly enforced discipline', he has said. The guerrillas, men and women, are forbidden to drink alcohol, they do not smoke and they refrain from sex. Their credo is to fight and sacrifice their life if necessary to achieve an independent state of Tamil Eelam. The focus of their loyalty, pending the achievement of the state, is Prabhakaran himself. Still only 36, he is regarded as almost a god and has the reputation of an incorruptible hero. His followers call him *Annai* or elder brother.

Without personal possessions except their AK-47s and a change of clothes, the Tigers live austerely. They are taught that the weapon is the most important object in their life and are urged never to let it touch the ground since 10 comrades could have died in the effort to capture it.

Anton Balasingham, a spokesman for LTTE, told an American journalist: 'We teach our fighters to transcend their egos and material pleasure, to subordinate their lives to a noble cause.'

It is virtually impossible for youths to reject the Tigers' 'invitation' to join their ranks as warriors. The standard first question posed to a potential recruit, male or female, is: 'Do you want to see our people free and in their own country or slaves under the Colombo regime?' The second question, usually addressed to a group, is: 'Hands up those who do not want to join the Tigers' student wing?' Of course, nobody dares to raise a hand. Through indoctrination, discipline, nationalism and leader-worship, the Tigers have become one of the finest fighting forces of modern times.

Same Battlefields, Same Hatreds

Some diplomats in Colombo predicted, late in 1992, that military exhaustion on both sides would bring the war to a 'dribbling end', as one described it to me. While there were periods when hostilities became more intermittent, the fear and hatred of each side for the other ensured a continuation of the fighting.

The biggest offensive for more than a year began in September 1993 when Sri Lankan government forces attacked rebel-held areas in the north. Soldiers, backed by air force bombers and strong artillery, thrust into the Jaffna peninsula in yet another attempt to cut a vital guerrilla supply route. In the first three days of fighting 175 rebels, 125 soldiers and many civilians were killed.

The troops had advanced 10 miles northwards from their base in Elephant Pass, where the Tamils were heavily entrenched. The army destroyed a well-fortified women's camp.

The objective was to seal off the Kilali lagoon on the peninsula, as the army has sealed off other routes and areas. Patrolling naval gunboats often clash with rebel boats making the crossing. Suicide Tigers in explosive-laden boats try to ram naval boats, sometimes succeeding. In August 1993 three naval craft were destroyed and 21 sailors killed in suicide attacks.

The lagoon is the only access the Tigers have to the mainland so they have no option but to fight to prevent it from becoming a government 'no-go' area. The offensive ground to a halt because the government was not prepared to accept the continuing heavy casualties. Once again Tamil ferocity and determination won for them a battle they should have lost.

During the latter part of 1993 and well into 1994 desultory fighting continued. The diplomats in Colombo were now saying that intermittent conflict could drag on for decades.

References

1. A British diplomat in Colombo has said: 'The victim's promotion will make the murderers feel even more pleased about their success. In this country of killers, to kill a General confers high prestige.'
2. Pararajasingham Joseph, a Tamil MP, is the source for those details. Foreign correspondents, diplomats and aid officials know Pararajasingham as a source of reliable information. He said that 47 people were killed inside a rice mill. 'Cannot somebody put a stop to this carnage?' he asked.
3. Elephant Pass is so named because elephants use the narrow causeway as a crossing. It is the army's last outpost in the interior of the northern province.
4. In October 1991 the Netherlands and Norway cut all aid to Sri Lanka because of human rights violations and the government's increased military spending.

War Annual No. 5 dealt with these aspects of the war: The training of women Tigers and their use in battle; the Indian Army's withdrawal from Sri Lanka; the Tamil offensive of 1990; the Amnesty Report on atrocities committed by both sides in the Sri Lankan civil war.

Sudan Civil War

BRIGADIER BASHIR'S 'FANTASY WORLD'

Background Summary

From 1955 to 1972, war between the Muslim Arabs of northern Sudan and the Christian negroes and animists of the south divided the country and prevented economic progress. A negotiated peace came into effect in 1972 but it proved meaningless and the state security forces continued to harass the Christians. In self-defence, they formed a guerrilla army known as *Anyanya* — venom of the viper — which grew into a conventional military force, the Sudan People's Liberation Army (SPLA). Colonel John Garang became its leader.

President Gaafer Nimeiri pursued a political role predicated to northern overlordship. Even when he was overthrown in a coup in 1985 the army continued to operate against the SPLA. The following year Sadiq al-Mahdi became Prime Minister and Garang confidently expected peace approaches but al-Mahdi, influenced by Islamic extremists, ordered the army to continue to war. Unable to bring the SPLA to pitched battle, the army massacred its civilian allies, the Dinka and Nuer tribes.

In 1988, Garang went on the offensive, moving northwards to capture the fortress town of Kapoeta. He held it until the Sudanese army had gone to immense expense to relieve the place and then he withdrew. Following this strategic victory, the SPLA inflicted several tactical defeats on the army.

In 1989 al-Mahdi moved closer politically and militarily to Libya and thus angered Egypt, whose President Mubarak considered Libya's Gaddafi a destabilising irritant. Al-Mahdi bought arms from Iran, a move which angered Saudi Arabia, and he maintained the harsh code of the *sharia,* the Islamic law, which disappointed and alienated potential Western allies.

On 30 June 1989, Brigadier Oman Hassan al-Bashir overthrew the al-Mahdi government and installed a 15-officer junta. Significantly, every one of them had been trained at Nasser Academy in Egypt and within hours the Egyptian government recognised the new 'Command Council of the National Salvation Revolution'.

Since Bashir, a devout Muslim, was intent on keeping the Christians of the south in subjection, John Garang did not anticipate any peace move from him and in fact he made none. Garang was in a strong position, with 50,000 men under arms, and he held the whole of southern Sudan east of the White Nile and south of the River Sobat.

Bashir, who was obsessed with the need to eliminate dissidents 'who infect the armed forces, the nation and the revolution', became ever more extreme. He called in Colonel Khair, who ranks third in the regime, to get rid of these supposed dissidents. On a single day Khair sacked 400 army officers, 380 police officers, 58 judges and many other officials. All were replaced with Islamic militants. Khair also abolished trade unions and professional groups.

On 23 April 1990, two of the dismissed officers, Major General Abdel Kader al-Kadro and Brigadier Muhammad Osman Hamad Karer, tried to overthrow Bashir in a coup. The coup was ill-timed, the plotters were arrested and 29 of them, including Kadro and Karer, were executed by firing squad.

Declaring *jihad* (holy war) against Sudan's Christians and pagans, Bashir found a way of conducting war against them on the cheap. He gave arms and ammunition to the northern nomadic Arab tribes and encouraged them to attack the southern peoples. A principal target were the Dinka people, who provided crucial support for Garang's SPLA.

During 1990, much of the war was fought along the roads, where the SPLA fired on army convoys on their way to relieve besieged garrisons. In one ambush in April more than 20 vehicles were destroyed when their cargo of ammunition, including heavy shells, came under fire and blew up. Three convoys failed to reach Yei, in the far south. Food is exploited as a weapon by both sides. The SPLA seizes food trucks for use by its own forces and both the SPLA and the army try to starve the garrisons of isolated posts into surrender.

Two weeks before Saddam Hussein sent the Iraqi army into Kuwait (on 2 August 1990), he resumed supplies of weapons and ammunition to Bashir's forces. He did this because Bashir supported him in his claims on Kuwaiti territory. Even while facing the threat of heavy attack by United Nations forces, Saddam maintained supplies to Sudan; the bombs used against SPLA-held towns in September and October came from Iraq. Since the general Arab policy was to oppose Iraq, Saudi Arabia was profoundly angered by Sudan's stance of support for Saddam and it cut off aid programmes to Sudan. Bashir's support for Saddam was obviously an error of judgement, indicating the Sudanese leader's political ineptitude.

The War in 1991

Bashir's rule in Sudan came under close international scrutiny in 1991 as a result of his support for Saddam Hussein and as his own people suffered grievously under famine. UN analysts described his style of government as 'groping, improvised and incoherent'.

It was clear from the outset of Bashir's administration that the survival of his government would depend on his success in dealing with the rebellious south and in coping with the corrupt system which al-Mahdi had allowed to strangle the country. As a staunch nationalist, and simplistic soldier, Bashir believed that he and his colleagues could solve their political and economic problems with a heavy hand and without an appeal to external assistance.

In all his first official statements, Bashir insisted that his government was neither Right-nor Left-wing and represented no tribal or ethnic interest. But many of his new ministers were either former members of the extremist National Islamic Front or they had close associations with it. After that, the Islamic nature of the government and of its plans for Sudan became ever more obvious.

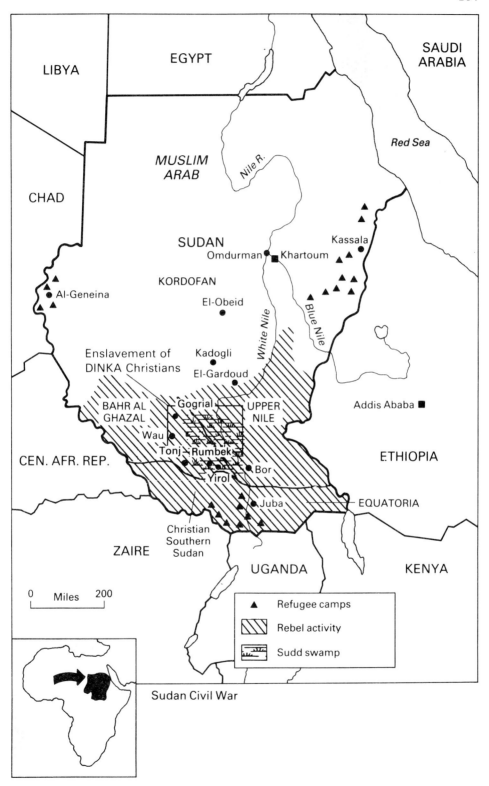
Sudan Civil War

Shortly after taking power, Bashir announced that talks would be held with the SPLA in Addis Ababa.[1] His lamentable ignorance on many serious issues became obvious at once. His fundamental approach towards the SPLA was 'we both fought in the desert and we understand each other. If we sit down together we will be able to resolve our problems'.

The SPLA, and Garang in particular, saw through Bashir. As a commander of troops fighting the SPLA, he had been crude and brutal in negotiations and had permitted his men to commit atrocities against SPLA soldiers and civilian populace. SPLA intelligence knew everything about Bashir and his genocidal outlook against the negro Christians.

Only weeks after his coup, Bashir's position in the civil war changed dramatically. He rejected the November 1988 peace accord, negotiated by the Democratic Unionist Party, and set about increasing conscription. At the same time he mustered all the army's resources for an even greater onslaught against the SPLA.

Sudan, he said, would push for 'union' agreements with Libya and Egypt. This upset the non-Arab southerners and it angered the Egyptian government, which does not like to be linked, even speculatively, with outcast Libya. Bashir proposed a referendum on the use of the *Sharia* but since 70 per cent of the populace is Muslim there could be no doubt about the result. If the proposal was an attempt to deceive a gullible outside world of his good intentions, it may have had a limited result. However, internally the suggestion was seen as yet another attempt to provide a false legality for a system which the Christians hate.

Other cosmetic steps Bashir has taken include the release of political prisoners, including Sadiq al-Mahdi; proposing a federal system for the country; and declaring a month-long amnesty during which SPLA rebels could give themselves up. Not a single SPLA fighter trusted Bashir sufficiently to come in on a promise of safe conduct.

The chaos surrounding the relief plan for the famine crisis is largely of Bashir's making. For more than a year he denied that a famine even existed. It was a 'Zionist plot' and an 'imperialist slander' even to suggest that the Sudanese people were hungry, he averred. Since there was no famine, he foreign aid agencies were not needed and the Sudanese ministries would no longer co-operate with them. At the time Sudanese were dying of hunger by the tens of thousands.[2]

Militarily, Bashir is regarded by only a few of his Sudanese contemporaries as able. He served with a Sudanese contingent during the Arab war against Israel in 1973 and he has had considerable experience in the deserts and mountains against the SPLA. Egyptian strategists and officers of Bashir's own rank do not rate him highly. An Egyptian analyst has said:

> Bashir's training as a paratrooper has given him the misconception that it is only necessary to drop enough paratroops on Garang's forces to unnerve them and cause a rout. If Bashir could indeed drop 50,000 soldiers they just might cause panic in the ranks of the SPLA. But his air force transport section has only one squadron with five C-130Hs, four C-212s, two DCH-5Ds and one F-27. It isn't even certain that any one of these aircraft is

equipped for paratroop drops. Bashir lives in a fantasy world.³

Bashir has no armed helicopters at his disposal and only one squadron of counter-insurgency aircraft. His strength, such as it is, lies in the army, which has 350 armoured personnel carriers, 240 towed artillery pieces and about 10 self-propelled 150-mm guns.

The army has not recovered from Colonel Khair's purge of officers in 1989. They were the most professional officers and their departure meant that the pressure against Garan's SPLA was instantly eased. Indeed, the sacking of these officers ensured that the army could not possibly defeat the SPLA. During 1991 Bashir asked his friend Colonel Gaddafi to send him some of Libya's best instructors and, in August, this request was met. The Libyans were not able to do much to strengthen the fighting ability of the army.

According to diplomats in Khartoum, Bashir is hoping that he can increase the intensity of his *jihad* against the Christians. He is hoping that if he can rouse his Muslims to carry out more massacres against the civilian Christians in their villages and on their farms then the SPLA fighters will desert in large numbers and return to their tribal areas.

It is rather more likely that such terror tactics will cause the SPLA to become even more determined to take the war to Khartoum and depose Bashir by force.

The War in 1993-94

During 1993 several intense efforts were made to persuade Bashir to give peace a chance. The presidents of Egypt, Kenya, Tunisia and Uganda visited Khartoum first, followed by the former president of the US, Jimmy Carter, and later the Archbishop of Canterbury. Bashir maintained the drive of his *jihad* and was urged on by discontented radicals from his National Islamic Front government who are frustrated by the continuing resistance of the Sudan's People's Liberation Army.

In January 1994 the Sudanese forces began an offensive which had been planned in November 1993. Thousands of troops, armoured personnel carriers and heavy weapons were sent by barge down the Nile from Khartoum. Their base was close to the southern city of Juba, about 120 miles from the Ugandan border. According to relief workers, Bashir was desperate and staking his future on defeating the rebels.

The government's first objective was to cut rebel supply lines by capturing the few remaining SPLA-held towns near the Ugandan and Zairian borders before the dry season ended at the end of April. Troops backed by heavy artillery and MiG fighter-bombers pounded SPLA positions on four separate sectors along an arc of 100 miles. The SPLA fell back under the sheer weight of the attack. Chief of Staff of the SPLA, Salva Kiir, said: 'They may take the towns but that will not end the war. We will take to the bush and defeat them from there.'

In the meantime the civilians, mostly Christians, suffered under the government onslaught. The town of Pageri, where there were no SPLA troops, was razed. Most foreign aid workers were evacuated from the area and 100,000 Sudanese dependent for their survival on outside help fled towards the

Ugandan border. The resumption of fighting came at a critical time for a region already suffering from the worst drought for years. Chief of the UN Operation Lifeline Sudan, Sally Burnheim, said: 'If the fighting goes on for three months more than four million people will be at risk from starvation.'

Diplomats and aid workers agree that the government's seizure of towns held by the SPLA would achieve nothing. For years, all foreign observers have known that the deep divisions between north and south, Arab and African, Muslim and Christian, cannot be solved militarily. During 1993-94 the crisis for the South was compounded by splits within the rebel movement that have aggravated tribal tensions and accelerated the carnage in a war that by March 1994 had cost the lives of 1.3 million Sudanese (UN estimate).

Late in February 1994, 12 relief groups working in Sudan urged donor countries to renew their efforts to induce Bashir to stop the offensive. He remained unmoved by their pleas. Helge Rohn of Norwegian People's Aid, in a broadcast from Addis Abbaba - she was forbidden to broadcast from Khartoum - said: 'There is no buffer for the people here - no cows, no grain, nothing whatever. If these people do not receive aid by April the dying will begin.' As she predicted, the dying *did* begin. Bashir's holy war grinds on.

References

1. Bashir was hoping to turn the Mengistu regime in Ethiopia against Garang. Mengistu had backed Garang because of their shared Christian conviction that Bashir, like Sudanese leaders before him, was intent on turning the Sudanese negroes into slaves. The overthrow of Mengistu in 1991 has not noticeably affected Ethiopia's support for Garang and the new rulers in Addis Ababa may be giving him even more practical support, in the form of captured Soviet-made arms, than Mengistu did.
2. Aid agencies say that Bashir makes their work more difficult than in any other country, with the possible exception of Afghanistan. 'Bashir would prefer to see his own people die than kept alive by Christian-donated food', a Norwegian aid worker told me. 'This indicates the extraordinary intensity of his Islamic belief.'
3. The Egyptian military analyst, M.E. Mahgrebi, considers Bashir the most incompetent military leader in Africa. He said, 'There are stories in the Egyptian army colleges that Bashir, as a captain and major, led his men disastrously and that he survived on at least three occasions by abandoning them and running away. His enemies would say this, naturally, but a confidential report by Egyptian military intelligence gives credence to these reports.'

Sumatra: Separatist War in Aceh

Aceh is the northern province of Sumatra, the largest island of Indonesia. Thousands of small islands off the Aceh coast are administratively part of the province, whose population is Muslim. A separatist movement, *Aceh Merdeka* or 'Free Aceh', has been fighting for independence since the middle of 1989 but information has been difficult to obtain and even more difficult to verify.

Ever since Indonesia itself gained independence from the Dutch after the Second World War, it has had problems with human rights. There was the massacre of probably one million people after a failed Communist-backed coup in 1965 and since 1975 the army has been engaged in a genocidal war in East Timor. The army tries to prevent information from filtering out of Aceh but occasionally a communiqué is issued about 'anti-bandit activities'. One such report, published in January 1991, stated that '20 to 30 bandits' had been killed.[1]

After criticism by the local press and international human rights organisations, notably the US-supported Asia Watch, the government announced in January 1991 that ringleaders of *Aceh Merdeka* would be put on trial. These trials, though largely held in secret, have produced some information about the insurgency movement.[2]

There is a tradition of rebellion in Aceh which began with an uprising against the Dutch colonialists in 1873. A 30-year war followed. Since 1949 sporadic uprisings against the Jakarta regime have been frequent. Over the decades, the Aceh fighters have won a reputation for fierceness and resilience.

The Aceh people are aggressively Islamic and are quite different from the more tolerant Muslims of Java. Javanese Islam has taken on many elements of Christianity, Hinduism and Buddhism.

Some evidence suggests that Islamic fundamentalists from Saudi Arabia and Iran have been sent to Aceh to turn it into a bastion of Islam in the South-East Asian archipelago.[3] Even stronger evidence establishes a link with Malaysia, only a short distance away across the Straits of Malacca. Malaysia gives sanctuary to the rebels actively being hunted by the Indonesian army in Aceh.[4]

The current rebellion began with the killing in June 1989 of Corporal Muhammad Gade, nicknamed 'the hunter' because of his skill in tracing rebel leaders and leading army units to them. The army killed some Aceh fighters, tortured others and in retaliation the freedom fighters struck at army posts and patrols.[5] The local commander, Major General H.R. Pramano, announced that he would crush the entire movement by the end of 1990 but this was a vain hope. He was forced to ask for thousands of extra troops to be flown in and the government in Jakarta officially classified the conflict as a war.[6] Numerous soldiers off duty were kidnapped by villagers and reportedly killed or mutilated. Their bodies were being found floating in the rivers and by the sides of roads.

Aceh Merdeka blames Indonesian atrocities for the continued fighting and some such atrocities are verifiable. For instance, in August 1990, Sulaiman Ampo Ali, a merchant from Matang Reubek in Aceh, was walking home when he was shouted at by soldiers. He kept on walking, only to be shot in the leg and thrown onto the soldiers' lorry. He was interrogated, taken to a nearby river, shot again and thrown into the river in a sack. Villagers witnessed much of what occurred and retrieved his body three days later.[7]

The army argues that force is necessary to prevent the Aceh population from breaking into warring factions. The roots of division are very evident, since hundreds of languages are spoken in Aceh, one for each island according to some scholars.

The army also says that the *Aceh Merdeka* movement is a front for marihuana producers who are prepared to pay the guerrillas to protect their lucrative trade. According to the Aceh people, the Indonesian army oppressively occupies the area, reflecting a punitive government policy. One element of this policy is to encourage businessmen from Java to dominate all economic activity in Aceh.

There is no doubt that a guerrilla war is under way. An Indonesian army doctor with service in Aceh and who has left his country recently, reported to human rights organisations that about 2,000 soldiers and rebels had been killed in the period July 1990 to July 1991.[8] According to his testimony the fighting is vicious on both sides. Hundreds of army patrols and naval patrols are in progress at any one time, while the Aceh resistance sets ambushes in the teeming jungles and along the rivers.

A report from a different source says that the Indonesian navy has upgraded its base at Saband, Sumatra, to deal with the *Aceh Merdeka* threat. About 20 inshore patrol boats, mostly of Yugoslav and Australian manufacture, have been deployed in Aceh waters. In addition, 4 Telul Langsa and 3 Teluk Semangka landing ships were moved to Sumatra harbours close to Aceh territory. The Indonesian navy is quite strong, with an establishment of 45,000, including 12,500 marines and an élite force of 1,000 naval air commandos. Some of the navy's ships were formerly part of the US, British and Dutch fleets.[9]

References

1. Indonesian army bulletin, 21 January 1991.
2. Patchy reports have appeared in the Jakarta press and according to Indonesian journalists they are censored.
3. This information comes from Iran itself. Some Islamic institutions in Tehran are remarkably frank about plans to spread Islam. Saudi Arabia and Iran, the bases for Sunni Islam and Shia Islam respectively, criticise Indonesia for not being devout enough in religious observance and they commend Malaysia for its more fundamentalist approach.
4. Information from diplomats in Kuala Lumpur.
5. Report from Asia Watch.
6. From a government document leaked to Asia Watch.
7. Asia Watch report from its own field workers.
8. The doctor, who now lives in the West, sent a report to the UN.
9. Information from officials of the International Institute for Strategic Studies.

Civil War in Former Yugoslavia

ANCIENT HATREDS, FRESH GRIEF
The Historical Background

The war in Yugoslavia has roots many centuries old. After the armies of each great European empire surged across the Balkans there remained a medley of Romans, Slavs, Turks, Habsburgs, Russians and Germans. In the 17th century, the Ottoman emperors drafted in large numbers of Serbs to settle along the border to serve as frontier guards. The descendants of that fighting people are now based in the south Croatian region of Krajina but they do not want to be in Croatia. Their desire is to remain in a unified Yugoslavia or, failing that, they want to be part of Serbia.

At the end of the First World War, the conquering powers took the remnants of Turkey's Ottoman Empire and the Austro–Hungarian Hapsburg Empire and turned them into the Kingdom of Serbs, Croats and Slovenes. Alexander I was placed on the throne as absolute ruler.

The new state contained an extraordinary collection of disparate and conflicting elements. The main ones were:

- Six republics — Serbia, Slovenia, Croatia, Bosnia-Herzegovina, Montenegro and Macedonia.
- Three religious groups — Eastern Orthodox Christian in Serbia; Roman Catholic in Slovenia and Croatia and Muslim in Macedonia. The other three republics had a mixture of religious beliefs.
- Several ethnic minorities, including Albanians, Greeks, Bulgarians and Hungarians.
- Two alphabets: Serbs use the Cyrillic alphabet while the Croats use the Roman alphabet. In addition at least seven languages are spoken in Yugoslavia.

From the beginning, the new state was a hotbed of nationalistic unrest, the result of centuries of rivalry, mistrust and hatreds. In 1928 Stefan Radio, Croatian leader of the powerful Peasant Party, was assassinated in the national parliament. His death at the hands of Serbian extremists set off bloody riots throughout Croatia. The following year King Alexander declared a royal dictatorship and tried to replace Serbian and Croatian nationalism with Yugoslav patriotism by changing the name of the kingdom to Yugoslavia, meaning Land of the Southern Slavs. He also made some constitutional concessions to the rebellious Croats.

A leading figure of the time was Ante Pavelic, a Zagreb lawyer, who created the *Utasha*, an extremist Fascist party. He went into exile in Italy where, supported by the dictator Benito Mussolini, he turned the *Utasha* into a terrorist movement that spread to Hungary and received much support from the new Nazi party. In October 1934 the *Utasha* Fascists, with the help of a Macedonian revolutionary, assassinated King Alexander and the French Foreign Minister, Jean-Louis Barthou, in Marseilles.

In the late 1930s the province of Croatia was created through an agreement between the Serbian government and the Croatian Peasant Party.

The outbreak of war in 1939 and the German–Italian invasion of Yugoslavia in 1941 allowed national hatreds full play and ended the monarchy. Hitler and his Nazi associates well understood the intense rivalries within Yugoslavia and saw the opportunity to win the Croats as allies. Hitler offered them, a people already bitter about Serbian dominance during the monarchy, their independence and nationhood.

Ante Pavelic was installed as Hitler's puppet ruler and launched a Nazi-type racist programme to 'purify' the new nation. One-third of Croatia's Serbian population would be forcibly converted to Catholicism, one-third expelled and one-third exterminated. Archbishop Stepinac and the other priestly Fascists actually sanctified this process. Just how many Serbians died in the *Utasha's* slaughter is not known precisely. The Serbs say 700,000, the Croats insist that it was 'no more' than 300,000. Other European historians say that at least 500,000 Serbs died.

Josip Broz Tito, who led Yugoslavia's fight for liberation from the Nazi occupation, was a Croat. Under him, the Yugoslav partisans, though largely armed by Britain, liberated their own country, unlike all the other European states, which had to depend on the US, Britain and the Soviet Union to free them from the Nazis. Tito (he became Marshal Tito in 1945) was respected by all Yugoslavs and it was his status, prestige and sheer strength of personality which held the fractious groups together. Tito was strong enough to break away from Moscow's domination and in 1953 he introduced a 'third path' between Stalinism and Western imperialism.

Tito created a collective leadership which was supposed to obviate any competition for supreme leadership among the various republics, but it turned out to be a weak system which relied on consensus. Under the fourth and final Yugoslav constitution, in 1974, the republics received the power of veto over central legislation. Immediately, the more prosperous northern republics of Slovenia and Croatia were pitted against the less prosperous southern republics. The northern republics resented having to pay for the development of the under-developed southern republics.

After Tito's death in 1980, the old cracks began to appear in the Yugoslav fabric. In particular, the smaller republics resented the central power of the Serbs, whose politicians made it clear that whatever system of government applied in Yugoslavia they would hold the reins of power.

The New Nationalism

Even so, Yugoslavia might have lived peaceably enough with its divisions had not Eastern Europe broken up during the astonishing events of 1989–90. The dictatorships of East Germany, Czechoslovakia, Romania and Bulgaria were overthrown and nationalism became the most powerful force in Europe. Inspired by the changes in the old Warsaw Pact countries, the Slovenes, Croats, Macedonians, as well as the Albanians in the autonomous province of Kosovo, held referendums and claimed their independence from Belgrade. Meanwhile Serbia was intent on building a greater Serbian nation, comprising Serbia itself, Montenegro, the autonomous region of Vojvodina, and Serbian enclaves in Croatia.

Croatia's crisis began in the late summer of 1990 when the Serb-populated areas in Croatia began to declare themselves autonomous, claiming that they had to defend themselves against what they called the 'oppressive *Utasha* regime' in Zagreb. In March 1991 Slobodan Milosevic, president of Serbia, began arming the Serbian population in Croatia, a highly provocative act. In retaliation, Croatia claimed a militant protective custody over areas in Serbia where Croatian groups were 'under threat'.

Croatia wants independence within its existing borders, which contain 600,000 ethnic Serbs. The Serbs in the countryside, supported by the Yugoslav People's Army (YPA) as well as the government of Serbia, do not want to be taken into an independent Croatia and militarily they have staked their claim to between one-fifth and one-third of the republic's territory.

On 25 June Slovenia declared its independence from the rest of Yugoslavia. Its people had long wanted to go their own way, largely as a result of their sense of closer cultural links with Western Europe than with the Balkans and the East. In 1991 it was by far the most prosperous of the Yugoslav republics and its citizens were protesting that their contributions to the federation's economy were being squandered and misappropriated.

Slovenia's forces consisted only of about 8,000 provincial police and 35,000 civilian militia but they had several factors working in their favour. For instance, they were fighting a guerrilla war in a mountainous terrain they know far better than their adversaries, with tactics designed by Tito himself. Alarmed by the Soviet invasion of Czechoslovakia in 1968, Tito had formed the territorial defence forces, lightly armed units whose task was to slow the progress of armoured columns of the very kind that the Yugoslav army sent into Slovenia on 26 June. The territorials' tactics, carried out with precision by the Slovenes, involved armoured columns being slowed down by tank traps and barricades and then attacked by swift mobile units.

The first real clash in the war may have occurred on 30 June at the village of Velika Vas in Slovenia. The No 2 Battery of the 580th Mechanised Artillery Brigade of the Yugoslav People's Army was ordered into action as part of a campaign to reassert federal control over the Slovenian republic. It was repeatedly held up by road blocks, which it cleared without much difficulty, until it found a barricade of buses barring the road at Velika Vas. Here it came under fire from Slovenian militiamen, who attacked with anti-tank weapons and rocket-propelled grenades. The YPA returned fire with anti-aircraft guns and called in an air strike. Two artillerymen died in the exchange and four were wounded — and the struggle for Yugoslavia had begun.

Elsewhere, lightly-armed Slovenian units struck hard against several federal detachments, which had not expected serious resistance. There was evidence that during these first clashes the army had acted of its own volition, without

206 THE WORLD IN CONFLICT

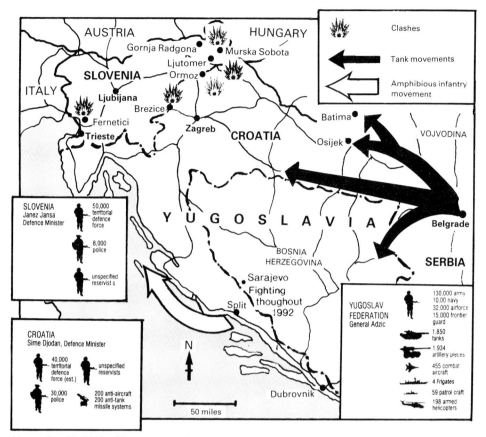

Yugoslav Civil War (August 1991)

authorisation from the Ministry of Defence in Belgrade. At the height of the fighting on 2 July, General Blagoje Adzic, the army chief of staff, appeared on television to issue a bellicose statement even as civilian officials were trying to negotiate a ceasefire. 'There is war in Slovenia', Adzic declared. 'A truce cannot be achieved so we will establish control and bring the fighting to an end.'

Throughout the summer of 1991 the Yugoslav army high command was cautious not to overstretch its resources, but the prospect of a war on several fronts became ever more alarming. By the end of October the army faced these problems:

- To establish a 'greater Serbia' the army needed to do more than hold the Serbian enclaves in Croatia. It had also to step in on behalf of the Serbs in Bosnia-Herzegovina, which had declared independence.
- A limited war in Bosnia to protect yet another 'endangered' Serbian enclave could easily turn into a religious war. The hatred between the Serb–Orthodox Christians and the majority Muslims runs even deeper than the animosity between Croat and Serb.
- The barracks in Belgrade were empty, as were the army's coffers. Without reserves of men and money it would be difficult to sustain the various campaigns.
- The Slovenian leaders, who had given a deadline (19 October 1991) for withdrawal of Yugoslav forces from their territory, stated that they would not allow further withdrawals after the deadline and would impound equipment as collateral against war reparations.

Mindless Struggles

The absurdity of the bloody war was shown by the struggle for Vukovar, between July and December 1991. A quaint town in the eastern Slavonian region of Croatia, Vukovar is in one of the two largest areas in the republic which are populated predominantly by Serbs. This gives it, at least to the Serbs, a significance disproportionate to its size and population — about 50,000. The mixed population had lived in harmony until Serbian propaganda convinced the Serbs that they were under threat. The Croats had done nothing more than dig in to demonstrate that they could hold out on their own soil.

The federal, largely Serbian, army intervened in force to demonstrate, for its part, that it could defend Serbs. It 'defended' Serbian civilians with such thorough mindlessness that barely a single house was left intact during the army's bombardment. Nevertheless, the ill-armed paramilitary forces fielded by the Croatians learnt that they could not stand up to the overwhelming superiority of the army. Casualties were heavy in this exercise in military futility.

The siege of Croatian-held Dubrovnik by the Serbs was an exercise in military spite. Considered the architectural jewel of the Adriatic, the ancient walled city came under siege on 1 October after desultory shelling. Dubrovnik has no strategic significance and the Serbs claimed it as their own, yet they relentlessly bombarded it as the 50,000 civilians huddled in cellars and shelters.

Occasionally the guns stopped firing while wounded civilians and 'monitors' sent by the European Community, were evacuated. Irreparable damage was done to the town and about 200 people were killed.[1]

Europe Fails the Test

The civil war in Yugoslavia was the first test of whether the nations of Europe could reshape a collective foreign policy now that Communism was no longer a threat and fast becoming an historical aberration. Here was a chance for Europe to prove that, in its newfound strength and prestige, it understood its political and moral obligations.

The scenario presented to the European statesmen was that, with the Iron Curtain torn down and the Soviet Union powerless to intervene, the Yugoslavs, still leaderless after the death of the giant Tito, were free to fight among themselves. The temptation was too much for the power-hungry men in Belgrade and they revealed themselves to be what Tito had always known them to be — Serbian Communist imperialists, determined to maintain control over their own republic and over others as well, notably Croatia, with its large Serbian population.

Governments across Europe condemned Belgrade for trying to carve greater Serbia out of neighbouring republics and for systematically destroying the civilian centres and cultural monuments of other nationalities, such as Dubrovnik.

The European Community agreed to apply economic sanctions against Yugoslavia, though they were aimed primarily at Serbia. One after another, in groups and singly, worried and well-meaning emissaries of peace arrived in Yugoslavia. Cease-fires came into force for a few hours and collapsed. A peace-keeping force was offered — but to become operational only after the fighting had stopped. As it did not stop, the EC sent in 'monitors', but all they monitored was the death of Yugoslavia and of thousands of people. In short, Europe was not up to the task of bringing about peace within its own borders.

Yet the only way in which effective intervention could come about was for the UN or the EC to send in a *peace-making* force. The EC would in fact be preferable because it could bring in the new democracies of eastern Europe and the Soviet Union, probably through the mechanism of the Conference on Security and Co-operation in Europe (CSCE) to help restore order. The CSCE's charter forbids the territorial expansionism that Serbia has been pursuing in its war against Croatia.

It would have been possible for the EC to announce three strategic objectives and, through the manpower and firepower available through NATO, to achieve them.

The first would be to drive the army, militias and national guards back to their bases and keep them there. The second would be to impose a truce during which negotiations could take place. Finally, the EC/NATO force would protect the international supervision of a settlement to guarantee the safety of all minorities within Yugoslavia, whether of Serbs living in Croatia or Croatians in Serbia or of other ethnic groups.

There is a precedent for such action in the new world order. When diplomatic and economic sanctions, as well as an aroused world opinion, did not persuade Saddam Hussein to leave Kuwait, a military expeditionary force was created to throw him out. It might be argued that Saddam had violated an international border while the Yugoslavian crisis and invasion was internal but internal squabblings can easily turn external and become international.

It takes only a little knowledge and understanding to see that four of Yugoslavia's neighbours — Romania, Bulgaria, Albania and Greece — could be drawn into the 'private' dispute. Also, refugees had started crossing the borders out of Yugoslavia within a few days of the fighting, thus internationalising the crisis. The danger of the war's spreading should surely be justification enough for an alarmed and worried Europe to take forceful action. The Americans need not have been involved militarily and their power was not required. The Yugoslav People's Army is much weaker than the Iraqi forces and most professional observers say that it would have put up no more than a symbolic resistance had an EC expeditionary *peace-making* force appeared at the Yugoslav border.

Despite Europe's indecisiveness, in mid-December 1991 the governments of Germany, Austria, Hungary, Denmark and Iceland announced that they would recognise the independence of Croatia and Slovenia, subject to certain conditions. Speaking after a long meeting of EC Foreign Ministers, the Prime Minister of Denmark said that the EC would invite the republics to ask for recognition of their independence claims by 23 December. This recognition would become official on 15 January, provided the applicants showed respect for democratic principles and human and minority rights. The Dutch president of the EC explained that delaying recognition until January would give the UN more time to get a peace-keeping force into Croatia and other republics.

Lord Carrington, who led the EC's largely ineffectual Yugoslav peace efforts, warned that premature recognition of Croatia and Slovenia could result in other republics being sucked into the war. There was a real danger that recognition of Croatia would prompt a demand for independence from Bosnia-Herzegovina, where strong forces of Serbian troops were stationed, and, indeed, this happened.

Nationalist Leaders

Slobodan Milosevic, President of Serbia. Aged 49, Milosevic was a law student and Communist Party functionary at the University of Belgrade in the 1960s and grew up under Tito's rule. An ultra-nationalist Stalinist, Milosevic became a banker by profession.

In 1984, as a fledgling politician, Milosevic was elected president of the Communist Party's Belgrade city committee. Just three years later, he rose to the republic's presidency by rousing Serbian nationalism over Kosovo, the southern province populated largely by ethnic Albanians to whom Tito had granted semi-autonomy. The Serbs have considered Kosovo the cradle of their civilisation since they lost a battle there in 1389. This defeat led to more than 500 years of brutal Turkish rule.

Considered a hard man by all observers, Milosevic exploits the dreadful behaviour of the Nazi Croatians in the Second World War to arouse fears among Serbs that history could repeat itself. Milosevic has the backing of the federal army, which is largely Serbian in composition, with a preponderance of Serbian Generals.

Milosevic has more personal power within his republic than any other Yugoslav leader yet, because of Serbian Belgrade's legacy as the country's intellectual centre, he faces a stronger opposition press than other leaders. The Serbian leader wins the support of Serbian nationalists by working on their age-old conviction that they are eternal victims, beset by enemies on all sides.

Milosevic brought about the crisis in the northern republic that led to war, then created a similar situation in Bosnia. Under his urging, four self-proclaimed 'Serbian Autonomous Regions' sprang up in Bosnia. One, in the north-west, lies alongside the main Serbian enclave in neighbouring Croatia and has declared that it will form a single entity. Another, in the south, enthusiastically sent reservists and volunteers to bombard the nearby Croatian city of Dubrovnik, which Milosevic claims is Serbian territory.

Franjo Tudjman, President of Croatia. A former army general, Tudjman carries the burden of being from the same community as the evil Fascist, Ante Pavelic, whose vile deeds are still held against all Croatians. There seems to be a hint of the Fascist in Tudjman himself, since in August 1991 he suspended democratic rights in Croatia and intimidated the press into self-censorship. He describes Croatia as 'a Christian wall against the infidels'. Tudjman's followers give him the straight-arm salute reminiscent of the Nazis and shout 'Franjo! Franjo!' in the way that Fascists applauded Hitler and Mussolini. Tudjman revels in the trappings of power, such as the guards of honour accorded to him on every official occasion and the parading of hundreds of colourful flags.

Just as Kosovo is sacred to Milosevic and the Serbs, so is the region of Krajina to Tudjman and the Croatians. A rocky area in south-eastern Croatia, Krajina is settled mostly by Serbs but has historic importance for the Croatians. It was in Knin, the declared 'capital' of the so-called Serbian Autonomous Region of Krajina, that the last Croatian king was crowned in the 11th century. Croatia might be prepared to offer the Serbian-populated regions some form of political and cultural autonomy but not the land itself.

Observers say that Tudjman is much less responsible than Milosevic for beginning the war in Yugoslavia. While he is an ardent Croatian nationalist, they say, he is more tolerant than his Serbian counterpart.[2]

The Yugoslav Army in June 1991

In June 1991, at the outbreak of the war, Yugoslavia had federal armed forces of 180,000 and reserves of 510,000. Of these the army consisted of 138,000 of which 93,000 were conscripts, with 440,000 reserves; the navy had a strength of 10,000, including 900 marines; and the air force strength was 32,000. In addition, Yugoslavia had 15,000 frontier guards. The Territorial Defence Force, a militia organisation, had a wartime establishment of 1.5

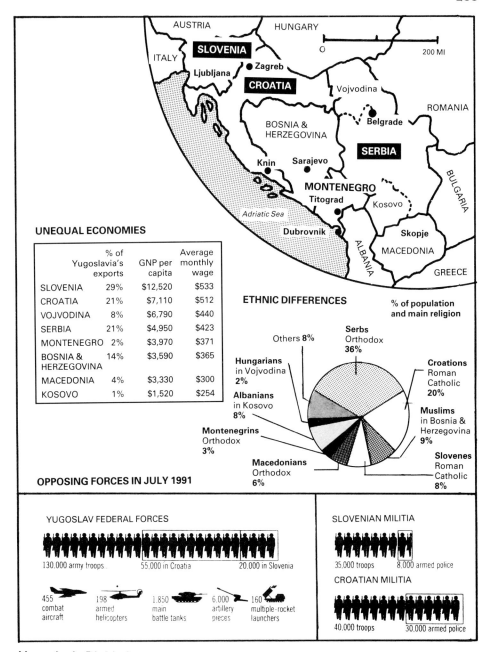

Yugoslavia Divided

million but only a fraction of this number was under arms, as a cadre, at the outbreak of the war.

The standing forces were well organised — the work of Marshal Tito — into four military regions, including a joint land-sea-air command for coastal defence.

In more detail, the army had 16 corps HQ, 3 infantry division HQ, 10 tank brigades, 6 mechanised brigades, 23 infantry brigades, 5 light infantry brigades, 1 mountain brigade, 14 field artillery brigades, 6 anti-tank regiments, 4 air-defence regiments, 1 airborne brigade and 6 SAM-6 regiments. The equipment of these units was modern, of high standard and of Western and old Eastern-bloc origin. Tanks alone numbered 1,900 but about 800 of them were in store. Artillery pieces exceeded 2,000.

Significantly, in view of the way the fighting developed, Yugoslav infantry has always been trained in street-fighting and counter-guerrilla operations. Some of the regular units were also specially trained in guerrilla fighting in case the country be overrun. This was another legacy from Tito, who had learnt at the outbreak of war against Germany and Italy that the Yugoslav army knew little about operating as guerrillas in mountain and forest.[3]

The strength of the air force, in relation to requirements for a civil war, lay in its 200 armed helicopters and its 12 squadrons of ground-attack aircraft, mostly Orao-2, Jastrab, Kraguj and Super Galeb.

The navy's strength lay in its frigates and its 59 patrol and coastal combat craft. It also possessed 35 amphibious craft, 10 of which could land tanks. The navy's bases are at Split, Pula, Sibenik, Kardeljev and Kotor.

The New War in Bosnia-Herzegovina

In the first week of April 1992, the European Community and the United States recognised the independence of Bosnia-Herzegovina but this move did not restore peace to the unhappy republic.

This area of the old, now defunct Yugoslavia, contains enclaves where Croats, Serbs and Muslims, respectively are the dominant ethnic groups. Croatia's president, Tudjman, and Serbia's Slobodan Milosevic, sent volunteers and weapons to aid their brethren in Bosnia-Herzegovina. For both these leaders, possession of Bosnian land was imperative in their conflict, which is aimed at incorporating the various enclaves of Croatian and Serbian people into their own states.

The complexities of nationalistic affiliations in the wreckage of Yugoslavia baffle many outsiders. They can be partly illustrated by the situation in the town of Listica in Bosnia-Herzegovina. An ethnically Croat town, it was originally known as Siroki Brijeg (meaning Wide Bank) but the Communists changed the name to Listica in 1945 because some of the worst pro-Nazi fascists came from Siroki Brijeg during the Second World War. As Bosnia-Herzegovina became independent, so did the local Croatian leaders revert to the name of Siroki Brijeg, largely because they are followers of Dobroslav Paraga, one of the most extreme Croatian nationalists. He has connections with the town.

In the town of Mostar, constant clashes take place between the Yugoslav Army and its allied Serb militants on one side and the allied Muslims and

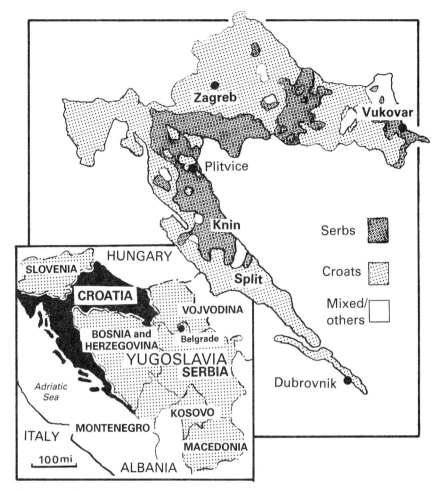

Croatia: Ethnic Groups

Croats on the other. The Muslim-Croat coalition accuses the army of being an 'occupying force' and demands that it withdraw. If this happened, the Croats and Muslims could more readily 'deal with' the Serbs.

In practical terms, western Herzegovina would be happy to secede from the new republic of Bosnia-Herzegovina; the people here see themselves as Croats first, Bosnians second. Elsewhere the reverse is true.

In April, three Bosnian towns, Foca, Bijeljina and Zvornik, as well as many villages, fell within a week to the Serbs and panic-stricken civilians in Sarajevo, the Bosnian capital, fought one another to flee the city before the expected full-scale Serbian attack. Bosnia's Muslims, the largest ethnic group, are poorly armed compared with the Serbs and have little tradition of fighting.

As in Croatia, the Serbian strategy in Bosnia-Herzegovina was to pick off one town after another. A leader known only as Arkan, who had become notorious in Croatia for his brutal operations, has been largely responsible for Serb activity in Bosnia-Herzegovina. Cleverly, he identifies the focus of Muslim resistance and hits it with an overwhelming attack. It was he who captured Zvornik, simply by hammering the defended police station with mortars and artillery and then overrunning it with infantry.

The president of Bosnia-Herzegovina, Alija Izetbegovic, accuses the Serbs of genocide against Muslims. He appealed to the world community, on 11 April, to prevent that genocide. However, the UN Protection Force (Unprofor) was unable to take any decisive action to help the Muslims. By mid-1992, it was clear that the UN Security Council must recognise the need for an enlargement of peacekeeping operations, so that Unprofor could deploy in Bosnia-Herzegovina — and before long in Macedonia and Kosovo as well.

In April 1992, Macedonia emerged as the new cockpit of Balkan intrigue. The EC had withheld recognition of Macedonia as an independent republic because of Greek objections. Greece accuses Macedonia of wanting to take over its northern Greek province of the same name. In spite of Greek objections, on 4 April, Germany established a consulate in the Macedonia capital, Skopje.

Recognition of Macedonia throws up problems other than that concerning Greece. The republic's population of two million contains a large ethnic Albanian minority which demands autonomy and, later, unification with Kosovo, the neighbouring Serbian-ruled province whose population is ninety per cent Albanian. Since Kosovo's leaders are in favour of merging with Albania, the situation could easily become similar to that which triggered two wars over Macedonia in 1912–13.

Many Serbs, like the Greeks, do not recognise Macedonians as a separate nationality and even talk of annexing the republic. However, Macedonians, Serbians and Greeks have one objective in common — they will fight fiercely to prevent the unification of all Albanians into one state.

The Conflict in 1993-94: 'Ethnic Cleansing'

Most of this period might be called 'the time of empty words'. The principal mediator (Lord Owen, who for some time operated with Cyrus Vance as his co-mediator), worked tirelessly and valiantly to find a peace formula but the belligerents, especially the Serbs, proved obdurate. Few international negotiators of any period could have laboured with such patience and skill as Owen showed. It was always going to be difficult to work with the Serbs, who advocated 'ethnic cleansing', one of the most odious terms in post-1945 politics. This meant the removal of non-Serbs from areas which the Serbs considered entirely theirs.

The main battlefield during 1993-94 was Bosnia and the main sufferers were the Muslims. The aggressors were the Bosnian Serbs, hell-bent on expanding the regions already under their control. The West seemed able to tolerate the final abandonment of Bosnia-Herzegovina, a state which the West had recognised, only to leave it to its fate. Over and over again, from 1992, the West and the UN promised and threatened but did nothing beyond declaring six areas to be safe zones. These were: Sarajevo, Gorazde, Bihac, Tuzla, Srebrenica and Zepce. Yet in mid-1993 the Bosnian government was being urged by the West to concede defeat by the Serbs for the sake of its own people. In effect, the West told the Bosnian leaders to accept a fragment of territory in which they could accommodate all those Bosnians who resisted the idea of an ethnically partitioned homeland or be ready to face the consequences.

By mid-1993 it had been demonstrated that appeasement had done nothing to reduce the bloodshed. 'At stake in Sarajevo', said the London *Independent* in an editorial on 26 July, 'is not just the life and health of its population, as though that were not enough ... it is the credibility of those flawed institutions that we have erected to fend off chaos. If the Western powers do not save Sarajevo they are saying by their inaction that they no longer respect the Charter of the United Nations.' And the editorial listed various other conventions, back to 1948, that were also no longer respected.

At that time the plight of Sarajevo (and indeed the other so-called safe areas) was desperate with shells raining on the city from Serb guns in the surrounding hills. People were dying as they queued for bread, children slaughtered as they played in the parks. The Serbs were relentlessly cruel.

On the morning of 12 February a mortar shell burst in Sarajevo's central market and killed 68 people. The horror of that event, seen on television worldwide, angered the Americans and Europeans who issued an ultimatum to the Serbs: 'Stop shelling Sarajevo. Pull back all the guns, heavy mortars and tanks 20 kilometres from the capital or put them under UN control.' This had to be done by 1 am local time on 21 February. Thereafter NATO warplanes would bomb or strafe any heavy weapons still in the exclusion zone or any guns firing on the city from beyond that zone.

The Serbs were not required to lift their siege of the city, nor were they forbidden to allow snipers to fire into Sarajevo. And there was no suggestion that foreign ground troops would be put into Bosnia to back up the UN ultimatum. The Serbs did make a show of pulling back and for a short time it seemed that peace might not be far off.

However, the military leader of the Bosnian Serbs, General Ratko Mladic, 53, is a man who cannot compromise, as his recent history shows. Assisted by the Yugoslav army and exploiting their overwhelming superiority in heavy weapons, forces under his command occupied a large swathe of Croatia and more than 70 per cent of Bosnia. Most of his victories came against hastily formed and lightly armed opponents. His offensives against Olovo, in January 1994, and against Maglaj in February, failed. In the 1991 war in Croatia he commanded Yugoslav Army forces in the heart of the Serb separatist rebellion.

He now talks of 'Serbian martial traditions', the 'Serbian national interest' and 'a unified Serbian state'. He is so powerful that he can defy with impunity his political superior, Radovan Karadzic.

In 1994 the Serbs' interest in possessing Gorazde in eastern Bosnia became critical. Their artillery pounded the city, whose population of 70,000 was swollen by refugees, and again NATO threatened to use its air power to force compliance with an order to withdraw. On 11 April NATO made its first attack against ground forces since the alliance was formed in 1950. The attack was ordered by the UN commander in Bosnia, Lieutenant General Sir Michael Rose, who feared that Gorazde was close to falling into Serb hands and called in two US Air Force F-16s. They bombed Serb positions in the mountains outside the city.

Rose reported that the shelling stopped less than 20 minutes after the bombing, but before long some of the most bitter and savage fighting of the Bosnia war took place and Serb forces burnt outlying villages which they had earlier seized.

The Serbs regard Gorazde as a strategic prize because of its proximity to the Serbian border and the threat it posed to them when in Muslim hands. The NATO/UN air strike did not do what the West intended. The Serbs' strategic thinking was superior to that of the West. They anticipated an air strike because it was by far the cheapest option for NATO. It would give President Clinton some brief glory, which he needed, but it would not cost American lives, which he could not afford. The Serbs were convinced that the West would not put into Bosnia the large numbers of troops needed to block and roll back the Serb advance. They also knew that even if the West were to stop the offensive against Gorazde, they could not prevent it from falling into Serb hands in the longer term, if only because the international humanitarian agencies would have to resettle the inhabitants of the town.

Confident of inactivity from the UN, NATO and the US, the Serbs returned to the attack during 16-17 April. They brought down a British Sea Harrier attempting a defensive bombing raid against them and they shot two British SAS soldiers near the front line, one of whom, Corporal Fergus Rennie, died. On 16 April, with General Mladic personally directing operations, columns of Serbian tanks, troops and trucks swept over the hills towards Gorazde. As the city came under heavy shelling, Serbian infantry moved in. The UN ordered new strikes and in poor weather two British Sea Harriers and two US 'tank busters' made three passes over the Serbian tanks but were unable to bomb. Elsewhere the Serbs were holding 200 observers of the UN Protection Force. The attempt to cow the Serbs ended abruptly.

The defiance of the Bosnian Serbs was a severe and possibly fatal blow to Western hopes of containing the conflict with limited military action. It was also a severe political and military humiliation, since the US had to turn to Russia to use its considerable influence with the Serbs in a desperate effort to restart peace negotiations and end the conflict. In Britain, the shooting down of the Sea Harrier and the death of Corporal Rennie brought renewed demands for stronger military action. In the House of Commons, the chairman of the all-party Bosnia committee, Patrick Cormack, said: 'It is absolutely essential that the United Nations sends troops to Bosnia. General

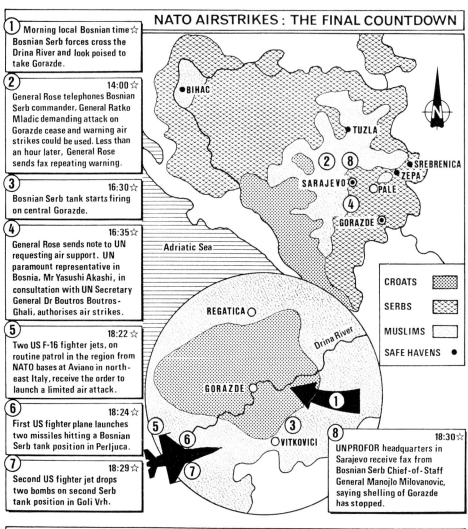
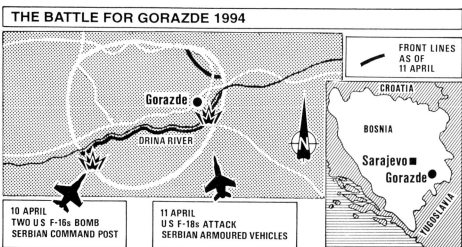

Rose quite manifestly is not getting the military power he needs.'

The commander of the land forces during the Falklands War, Major General Sir Jeremy Moore, said that it was 'exceedingly difficult' for military commanders when they were not given the freedom of action or the firepower they needed.

The West launched its air strike much too late and apparently had not considered what it would do if the Serbs defied it. The Balkan conflict could become a war by proxy between East and West. Russia has always opposed air strikes partly because of an historic affinity with the Serbs and because Moscow is unwilling to allow NATO the right to intervene in the old Soviet 'sphere of influence', which Russia still claims. When NATO threatened to use force to lift the siege of Sarajevo earlier in 1994 Russia sent its own troops to take up positions around the capital. The West cannot make air strikes or engage in ground action without consulting Russia.

Muslims and Croats

The US initiative to stop the fighting between Muslim Bosnian government forces and the HVO, the Bosnian Croat militia, was more successful than peace initiatives elsewhere. A fragile peace developed in March 1994, along the 100-mile Muslim-Croat front lines in central Bosnia and Herzegovina. The fighting ended because the Americans, through their peace envoy Charles Redman, convinced the Croats that peace was in their interest.

Croatian president Franjo Tudjman had long supported the Croats of western Herzegovina in their attempts to carve out a mini-state within Bosnia. This was disastrous because it forced the Croats into an uncomfortable alliance with Serbs and imperilled the 60 per cent of Bosnian Croats who live outside Herzegovina, among Muslims.

Troops from Britain, Ukraine, France and Turkey patrol the Croat-Muslim ceasefire lines. The centres relieved of fighting were Zepce, Stari Vitez and East Mostar. However, given the instability of former Yugoslavia and the extraordinary nature of changing alliances, long-term peace cannot be guaranteed.

References

1. According to International Red Cross estimates 7,500 people, mostly civilians, had been killed and 20,000 wounded up to 31 December 1991. Some journalists speculate on a much higher figure as a result of massacres, for which little reliable information is available. The economy of Yugoslavia as a nation and of the individual republics was in ruins by December.
2. According to the former British Prime Minister, Margaret Thatcher: 'This is not just a struggle between national groups. It is between Communists and those — the Croats — who seek democracy.' Speech in Parliament, 1 October 1991.
3. An American journalist, Michael S. Serrill, quotes a senior US intelligence source: 'The Yugoslav army isn't worth a dime. They've never fought a war. Most of the ranks are conscripts and the officers have divided loyalties. It's not a crowd to be counted on.' *Time Magazine*, 15 July 1991.

War Trends — Racing Towards Conflict

The word **trends**, which implies gradual or even slow movement, might not be appropriate to describe the rapidity with which governments, nations, communities and peoples moved towards conflict in 1991 and 1992. The rate of political change is breath-taking and in no case, anywhere in the world, is the risk of war less because of it. It has been frequently necessary for statesmen of the responsible, more stable nations to move quickly in order to prevent crises from becoming disasters; not that they always moved quickly enough, as the onset of the Yugoslav civil war proves.

In the first week of December 1991 the announcement that a commonwealth of independent states (CIS) had been formed by republics of the old Soviet Union came as a surprise to the Bush administration. The President of the former Russian Federation, Boris Yeltsin, telephoned President Bush with the information only hours before the public announcement of the new union. At the time, Western leaders were still of the view that a continuing central authority in Moscow under Mikhail Gorbachev would be the best hope of the Soviet Union's extricating itself from political and economic crisis. But there *was* no Soviet Union and Gorbachev was a phantom leader in charge of nothing more than a detested memory.

Analysts of the Soviet Union in the US could barely keep abreast of developments. The former national security adviser, Zbigniew Brzezinski, said on 9 December that the day was 'a turning point in world history', and that the world was 'on the brink of momentous happenings'. Only six weeks before Brzezinski's comment, the US Secretary of State, James Baker, had gone to Moscow to talk about conditions in the Soviet Union after the attempted coup against Gorbachev and had come away insisting that Gorbachev's central authority could continue. The Russian republic under Yeltsin, said Mr. Baker, had plans to deal with the nuclear weapons in the various republics by bringing them all back under Russian control. Baker actually believed that this would happen. He was mistaken and, instead, Russia, Ukraine, Byelorussia and Kazakhstan each indicated that they would retain their nuclear weapons. Baker intended to suggest that the Soviet republics use the framework developed by NATO for control of nuclear weapons as a guide. The NATO system is for central control but involving representatives of all member nations. The Secretary of State did not, at that time, get the opportunity to make this suggestion because he was overtaken by events.

It was the Ukraine's overwhelming vote for independence that finally drove home to President Bush the fact that the Soviet Union had ceased to exist. In an intense flurry of activity in the first week of December, the CIA set to work to prepare eight separate studies but was already making a preparatory prediction that catastrophe was looming. The former Soviet Union was sliding into chaos, with the imminent collapse of manufacturing and food distribution and the danger of hyper-inflation of more than 1,000 per cent a year. Civil war was a possibility, even a likelihood. The former Soviet Union could become a 'nuclear Yugoslavia', the CIA warned.

The State Department had already produced a report indicating that the future of democracy in the republics was uncertain, pointing out that Hitler had risen to power on the back of massive unemployment, hyper-inflation and lack of central control.

President Bush ordered Robert Gates, newly-appointed as CIA chief, to sharpen US intelligence capability. It was essential for the CIA to be able to predict crises and their probable results, particularly in the area covered by the old Soviet Union. Gates immediately responded by informing the President that Gorbachev could be written off; he could play no further part in what was happening to his former empire.

In fact, Transcaucasia was already on the brink of war. Presidents of the republics, such as Levon Ter-Oetrosyan of Armenia and Ayaz Mutalibov of Azerbaijan, had no control over the guerrillas fighting about the disputed territory of Nagorno-Karabakh, where more than 1,000 people were killed in 1989–91. Conflict over Nagorno-Karabakh, which is claimed by both republics, worsened after a helicopter carrying peace mediators crashed in the enclave in December, killing all the passengers. Armenia said that the helicopter crashed because of stormy weather but Azerbaijan claimed that it was shot down by Armenian militants.

Unrest in the Soviet armed forces is a matter of great concern to the West. A conference in Moscow involving high-ranking Soviet and NATO military officers was held in an atmosphere of near despair in December 1991. Liberal Soviet officers complained that the Defence Ministry hierarchy had failed to sweep away all the old guard command structure that had been imposed by Marshal Dmitri Yazov, the disgraced former defence minister. They reported that, while the new army chief of staff, Army General Vladimir Lobov, had replaced the senior officers, whole layers of Yazov appointments lower down the scale were blocking reforms. Without reforms, coups could occur within the armed forces.

About a week later, Defence Minister Shaposhnikov dismissed Lobov and replaced him with General Victor Samsonov, adding to the uncertainty in the West over who had command of the arsenal of 27,000 nuclear warheads. Following the August coup against Gorbachev, it is still not clear whether there is a KGB element in the nuclear chain of command.

The Islamic Bomb

The present or future President of the US could face a situation in which a group of Islamic nations controls a nuclear arsenal capable of producing a holocaust. A report by the CIA in December 1991 suggests that a commonwealth of Islamic Soviet states could present 'a grave danger' to the Middle East and to international strategic politics. It would be based on Kazakhstan's control of several thousand nuclear warheads. While the Muslims of that former Soviet republic are not as overtly militant as those in Iran, they are under pressure from the Iranians, who are extending their influence into other Muslim countries.

Western statesmen hope that Kazakhstan's President, Nursultan Nazarbayev, will sign the Strategic Arms Reduction Treaty and accept the nuclear reductions promised in August by Mikhail Gorbachev, even though he is not bound by such commitments after the collapse of central authority.

Under the greater religious tolerance introduced by President Gorbachev, Islamic fundamentalism has flourished in the Muslim republics of the old Soviet Union. Hundreds of new mosques opened during 1991 and, partly as a result, Islamic fundamentalists are well armed, highly secretive and insistent on an Islamic state. Neither Communism nor any reform movement can provide salvation for Muslims, they say. Islam must rule Muslim states. It is of some consolation to those who fear the march of Islamic fundamentalism that in the former Soviet Islamic republics there is no monolithic Muslim party; as always, Muslims are divided along ethnic and sectarian lines.

However, radical Shias in Iran, the Mujahideen in Afghanistan and the extremist *Jamaat-a-Islamie* in Pakistan are bent on creating chaos in central Asia because they know that it will lead to greater opportunities to spread Islamic fundamentalism. They fund Islamic parties and circulate literature — much of it inflammatory. The Afghan Mujahideen smuggle arms into Tadzhikistan and Uzbekistan.

The leadership in the five Muslim central Asian republics — Kazakhstan, Uzbekistan, Tadzhikistan, Kirghizia and Turkmenia — face increasingly hostile threats to their authority from the dozens of underground Islamic parties that have sprung up in their republics.

Quite apart from the danger posed by nuclear weapons in the hands of Islamic zealots in the former Soviet Union, a great drive is underway by certain Islamic nations to acquire a Muslim bomb. It was ironic in 1991–92 to see the Middle East and countries peripheral to it such as Pakistan, intent on obtaining nuclear weapons even as President Bush was announcing radical cuts in the US nuclear arsenal.

While international attention was almost wholly fixed on the search for nuclear facilities in Iraq, four other countries — Algeria, Syria, Libya and Iran — were rapidly developing their nuclear programmes. At the end of 1991 perhaps the greatest immediate threat was posed by the regime of the Chadli regime in Algeria, which was actively striving to acquire an 'Islamic bomb'. President Chadli happened to be one of the closest political allies of Saddam Hussein of Iraq.[1] Chadli's successor, Muhammad Boudiaf, is following Chadli's policies.

The President of Libya, Colonel Muammar Gaddafi, has had a relatively low-key nuclear research programme in place since the 1970s. Following the revelations about the advanced state of Iraq's nuclear development, Gaddafi launched a serious and intensive effort to catch up with Iraq. Almost certainly his intelligence system did not know that Algeria was well on the way to being able to produce plutonium. When this information reached him in 1991, he announced to his associates that Libya would not be left behind in the Middle East nuclear arms race.

His priority is to create a uranium enrichment facility based on electro-magnetic technology. Transfer of information about this system is not hindered by any category of international control mechanisms.

Syria is further advanced than Libya because the Soviet Union set up a nuclear research reactor in Syria during its years as Syria's protector and supplier. Syria has firm contacts with Western industrial concerns for the building of nuclear power stations.

The President of Iran, Ali Akbar Rafsanjani, is another Muslim leader who was profoundly worried about the development of Iraq's nuclear capability, when this became common knowledge after the end of the war against Saddam. Iran was already working on a nuclear programme, using a reactor built with Chinese assistance. With new impetus, Rafsanjani created a special ministry for nuclear development. This body put out feelers in various directions and quickly established contacts with a number of countries, among them Brazil and Argentina, both in the front rank of emerging nuclear states.

Pakistan, which in the military sense is the leading Islamic nation, is believed to have 10 nuclear devices ready to be assembled. For several years Pakistan and Libya were closely connected in the development of an Islamic bomb and other Muslim nations would naturally look to Pakistan for advice and practical assistance. A nuclear arms objective financed by oil revenue from the Gulf must succeed. Israel, as the prime target for most of the nuclear-hungry states, has ever need to get its intelligence predictions correct. Mossad, its external intelligence service, repeatedly reports that Islamic states will use nuclear weapons 'during the 1990s'. While Israel is reported to have up to 100 nuclear warheads, it will have great difficulty in competing with the wealthy Arab nations.

The acquisition of nuclear weapons has become a prestige symbol in the developing world generally and in the Muslim world in particular. In elections during December 1991 and January 1992, Islamic fundamentalists in Algeria had apparently won control of parliament, an alarming development because the Islamic National Front in Algeria has frequently declared its interest in creating an Islamic state to lead the rest of the Arab world 'in every way — morally, politically, militarily'. The fundamentalists' slogan is 'No constitution and no law. The only rule is the Koran and the law of Allah'. Under the Koran, all non-Islamic states must, in time, be converted to Islam by one means or another. In the event, the elections were suspended.

Meanwhile, nuclear arms development in isolated Communist North Korea is the most serious threat in Asia. Intelligence reports indicate that North Korea could produce such weapons by 1993. In December 1991 the country's leadership said that it would sign nuclear safeguard accords, but not until the withdrawal of US nuclear arms from South Korea was verified. Such a signing is meaningless without nuclear inspections, which North Korea rejects. Since there is incontrovertible evidence that two of the three nuclear reactors are operational and that work to develop reprocessing technology is well advanced, UN inspections could be too late.[2]

Syria and the North Korean Scuds

In March 1991, Syria became a renewed threat to peace in the Middle East when a North Korean freighter docked in Latakia and unloaded 24 medium-range Scud missiles and 20 mobile launchers. The missiles, code-named Scud-Cs by Israeli military analysts, have a range of about 300 miles, nearly double that of the standard Scud-B missiles, which had previously been the longest range missiles in Syria's army. The Scud-C's greater range put every town and settlement in Israel within striking distance of Syria. The Scuds which Saddam Hussein fired at Israel could not reach much further south than Tel Aviv.

The source of the Scuds, North Korea, is alarming. The government in Pyongyang has built up a formidable arms industry. A number of Third World governments, mainly in Asia and Africa, have bought tanks, jet aircraft, patrol boats, artillery, armoured personnel carriers and ammunition from North Korea.

From the outbreak of the Iran-Iraq War in 1980 until mid-1992, North Korea was a major supplier of arms to Iraq. Then it abruptly switched its support to Iran, perhaps because Iran promised regular supplies of oil. Whatever the reason for the change of client, the Iranian offer was obviously one that North Korea could not refuse, since it meant losing $1 billion owed by Iraq. Between mid-1982 and August 1988, North Korea sold Iran at least $5.5 billion worth of weaponry, making it easily the biggest supplier among non-Soviet Communist countries. By 1984, North Korean military advisers were helping Iran's war effort.

The missile deal with Syria may have been made before the War against Saddam began. For four years Syria had been trying to buy accurate surface-to-surface missiles, such as the SS-23, from the Soviet Union but was repeatedly denied such a deal.

The quality of the new North Korean Scuds is uncertain. They may be an upgraded version of the older Scud-B that the North Koreans copied through a process known as reverse engineering. Equally, they may be a newer, better version of a copy known as the Scud-PIP (for Product Improvement Programme).

President Hafez al-Assad of Syria and his Foreign Minister, Farouk Shara, told US Secretary of State James Baker that the arrival of the Scud-Cs should not be seen as sinister, since they had been ordered so long ago. Syria, they stressed, was emphatically a partner of the US and its efforts to resolve the dispute between Israel and the Arab states. But this provided no reassurance, since Syria's objective of crushing Israel has never changed; the time at which the Scuds were ordered is an irrelevance. There were rumours in international arms circles, late in 1991, that other Scud-Cs had arrived at Latakia.

Little publicity has been given to North Korea's assistance to Egypt to build a Scud-B plant. This became effective in 1987 and the plant is now operational. North Korea itself builds about 70 Scuds a year.

Instability in Africa

Africa remains a volatile continent, as was demonstrated during 1991–92 by riots which did not quite reach the 'status' of war. In Zaire, in September 1991,

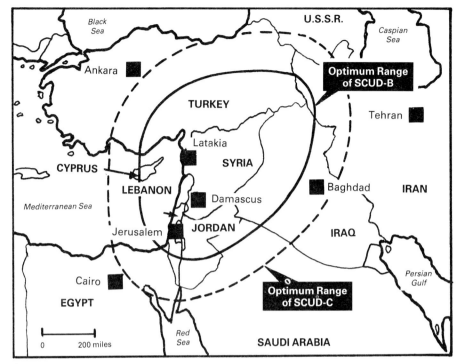

Syria's North Korean Missiles

a few hundred paratroopers protesting about low pay started a conflict that challenged the dictatorship of President Mobutu. The rebellion began when troops stormed out of barracks near Ndjili airport and pillaged Kinshasa as if it were a foreign capital captured after bitter fighting. The rebellion quickly spread to other military bases and cities.

Many foreign residents, including 40,000 Portuguese, locked themselves in their homes or sought refuge in their embassies. To protect and evacuate their nationals, France and Belgium flew in troops, who took control of downtown Kinshasa and the airport. In the midst of the crisis Mobutu found himself deserted by former allies in the West. Belgium, the US and France, which once supported his regime as a bulwark against the spread of Soviet influence, now saw his corrupt despotism and bankrupt economy as embarrassing and destabilising.

After more than 300 people had died in the violence, with 2,000 injured, Mobutu promised to share power with the opposition and, in a show of strength, demanded that all Belgian troops in Zaire should be withdrawn at once. The Belgian government rebuffed him insisting that its troops would pull out only after the evacuation of all Belgian nationals was completed.

If Zaire were to collapse, the horrors which engulfed Liberia in 1990 could be repeated. Mobutu took the precaution of ordering his 81,000 soldiers, 6,500 members of the 'Presidential Division' and the 12,000 members of the civil guard not to harm foreigners.

While Mobutu wanted to get foreign troops out of his country, the Prime Minister of **Togo**, Joseph Kokou Koffigoh, appealed to France on 30 November 1991 for armed intervention in the face of an attempt by rebellious troops to bring down his civilian government.

Troops loyal to the President, Gnassingbe Eyadema, who had been stripped of most of his powers in August, ordered an overnight curfew and threatened to reduce the capital, Lomé, to ashes if a new government was not named immediately. Rebellious army units continued to surround government headquarters where Koffigoh and the French ambassador were trapped. Koffigoh's transitional government is supposed to return Togo to democratic rule but, as in many parts of Africa, many factions and tribes in Togo see no virtue in democracy.

Also in 1991, the army of **Uganda** killed about 500 rebel guerrillas in operations in northern and eastern parts of the country. A range of rebel groups, some supporting former presidents Idi Amin and Milton Obote, have been trying to overthrow President Yoweri Museveni since he took power in a military coup in January 1986. The anti-rebel operations were linked to a wave of arrests, including prominent politicians. Those arrested included Omara Atubo, Minister of State for Foreign Affairs, and Andrew Admiola, former Ugandan High Commissioner in London.

In **Madagascar**, the President, Didier Ratsiraka, declared a state of emergency in the capital, Antananarivo, sending armoured cars into the city centre and ringing government offices with guards. The opposition, the 'Active Forces Coalition', appointed a retired general, Jean Rakotoharison, as 'president' of an alternative government.

President Ratsiraka, a former radical naval officer who took charge of the

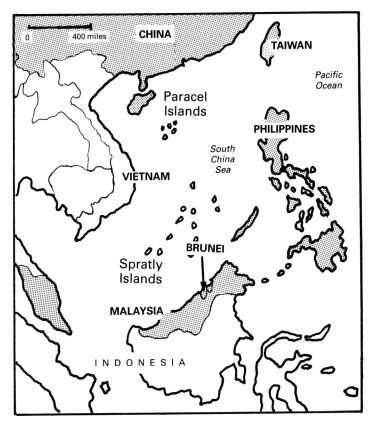

Spratly Islands and the six nations which claim them.(China and Vietnam also dispute the Paracel Islands).

country's ruling military junta in 1975, has switched from Marxist to market economies in recent years and has introduced some slight democratic reforms. Opponents accuse him of rigging elections in 1989 which gave him a third seven-year term in office.

Conflict — almost certainly war — looms over possession of the Spratly Islands, in the South China Sea, 160 miles north-west of the Malaysian state of Sabah. The islands consist of 53 insignificant reefs, shoals, cays and islands. But they are believed to have immense resources of oil under them. During this century six nations — China, Taiwan, Vietnam, Malaysia, Brunei and the Philippines — have claimed ownership of the islands. Since 1980, the Spratlys have been covered by a blanket of military security imposed by each and every nation claiming sovereignty over particular islands and areas of sea. They are out of bounds to everybody except local fishermen — and even their activities are monitored.

China has intensified its naval presence in the area in reaction to what a spokesman called 'the intensified scramble for all resources after the Gulf War'. Hostility between China and Vietnam is leading to expanded militarisation of the area. Malaysia is attempting to turn Swallow Reef into a tourist site for scuba divers and deep-sea fishermen, but this is a transparent attempt to extend Malaysia's territorial limits further into the South China Sea. The Philippines government is engaged on a similar exercise.

Sadly, but not surprisingly, many governments, societies and communities, as well as ethnic and religious groups, find it easier to turn to force to settle a problem or grievance rather than engage in the laborious process of negotiation. Negotiation and its concomitant, compromise, are perhaps too intellectual a means to resolve difficulties.

This certainly seemed to be the case on 22 June 1992 when the President of Moldova (the former Moldavia) announced that his republic was at war with Russia. President Mircea Snegur's statement followed a weekend of fighting in the conflict between ethnic Romanians and Russian speakers in Moldova. Russia's 14th Army lent its equipment and also some of its soldiers to the cause of the Russian-speaking separatists in the self-proclaimed republic of Transdnestr.

President Snegur said that he considered this territory, which is east of the Dnestr river, to be under Russian occupation. Between 19–21 June Moldovan forces captured the city of Bendery from separatist control in an offensive spearheaded by tanks. The Moldovans then lost the city again. Casualties in this and other fighting amounted to perhaps 2,000.

Most of Moldova was annexed from Romania in 1940, and 60 per cent of the population are ethnic Romanians. However, the east bank territory is mainly Russian-speaking and was never part of Romania. The separatists are convinced that Moldova plans to reunify with Romania, making Russians and Ukrainians inferior citizens. This is in only one war of several that could erupt before the next edition of *War Annual*.

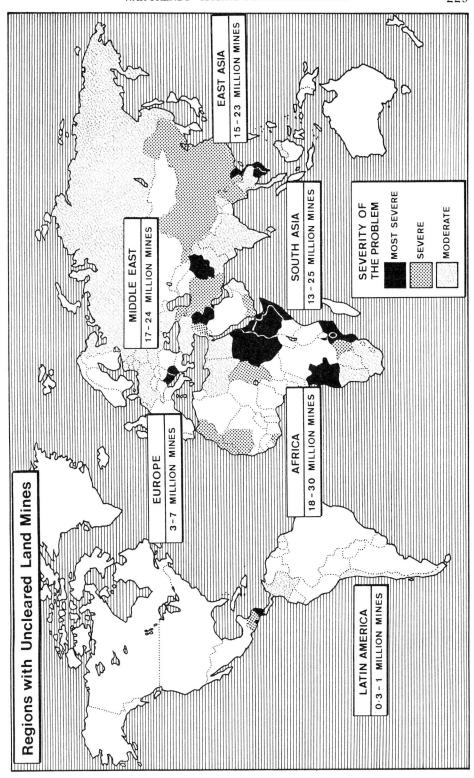

Rwanda's Machete War

On 6 April 1994 the presidents of Rwanda and Burundi were killed when a missile brought down the aircraft in which they were travelling. This assassination ended a fragile truce between the Hutu and Tutsi and a ferocious war began. President Juvenal Habyarimana, dictator of Rwanda for 20 years, was a Hutu and his tribe blamed the Tutsis for his death.

Periodically the Tutsis have massacred Hutus in their thousands. Since the middle of 1993 the Hutus have turned on their former masters. Nothing can explain the intensity of hatred one group feels for the other, especially as the two are inextricably linked by history, culture and language. The two groups have mixed so much that it is difficult for outsiders to tell the difference. The gangs of killers armed with machetes and sharpened bamboo spears thought they could do so and they slaughtered all those who looked like 'the enemy'.

Rwanda has two armies: the government army, commanded by General Augustin Bizimungu since the death of General Gatsinzi, who was also killed in the plane crash, and the army of the Rwanda Patriotic Front (RPF) of the Tutsi tribe which attacked the barracks of the Hutu-dominated presidential guard in the capital city, Kigali.

The RPF managed to get a battalion of troops into Kigali, where they became locked in a struggle for its control. From their stronghold in the north of the country a column of about 1,000 more RPF troops linked with those in Kigali.

As always in an African country, the victims were the ordinary people. First the Tutsis were hacked to death by the presidential guard in reprisal for the president's assassination. Then the Tutsis set about the massacre of Hutus. This mass madness led to the deaths of tens of thousands of people, according to the International Red Cross in Geneva. At least as many were wounded. In Butare, Rwanda's second city, 60 miles south of Kigali, gangs of men set fire to settlements and hacked residents to death.

With difficulty, French and Belgian paratroopers evacuated nearly all the several hundred foreigners in Kigali but the foreign troops would themselves have been butchered had they intervened to save the lives of Rwandans. The war ended indecisively and further bloodshed is inevitable in a society in which hatred and a desire for vengeance are the dominant feelings.

Gangs from Rwanda sometimes sweep into neighbouring **Burundi**, where many Hutus are refugees. In February 1994 Tutsi soldiers from Rwanda swept into some Hutu villages and butchered hundreds of people for no other reason than that they hate the Hutus.

Burundi is the scene of endemic fighting. In October 1993 Tutsis, who dominate the Burundi army, overthrew and killed President Melchoir Ndadaye. He had been in power for only three months. The Tutsi coup collapsed and its leaders fled from the country. The government had also fled to Rwanda where it established itself as a government-in-exile. Yet again, the ordinary people suffered as gangs butchered people indiscriminately. Thousands died and an estimated 600,000, one tenth of the country's population, fled across neighbouring borders.

Four times since independence from Belgium in 1962 the Hutus and Tutsis have waged war against each other. Much of the massive killing was blamed on the minority Tutsi, who make up 15 per cent of the population but control the government, most of the economy and, crucially, the military.

Like many countries in Africa, Burundi had been attempting a transition to democracy and in June 1993 had elected its first president since independence. Ndadaye, a Rwandan-educated banker, was a symbol of a new age dawning for the Hutu underclass. His successor, Cyprien Ntaryamira, was killed with Juvenal Habyarimana when their plane was shot down. Two presidential deaths by assassination have ended Burundi's democratic experiment and they set a depressing example for democracy in Africa as a whole. Tribalism is rampant and ethnic hatreds are feeding it.

In West Africa, **Liberia** is still suffering from the effects of the civil war following a ceasefire in July 1993. The country has a confusing pattern of front lines and no-go areas. Charles Taylor's National Patriotic Front of Liberia claims to be the protector of national sovereignty against 'foreign aggression' by the Nigerian-led West African Peace-keeping Force. The Front holds two-thirds of Liberia. The remainder is controlled by another armed faction, the Armed Forces of Liberia, nominally under the control of the ineffectual interim government of Monrovia. Either of these well-armed factions is capable of renewing the war and each periodically carries out massacres. The National Front killed 400 refugees, including 103 infants, because they were receiving rice rations when the troops were not.

In the face of so much violence, the United Nations, short of money, short of manpower and criticised for its performances, has reached the end of its capacity for settling global disputes. In 1988, when UN peace-keepers won the Nobel Peace Prize, their number totalled 10,000. In 1994 almost 100,000 blue helmets are deployed around the world. Everywhere the UN has achieved only part-peace. Troops still patrol truce lines but now they also monitor elections, protect human rights, train local police, guard humanitarian relief deliveries and, most recently, use force against those who get in their way. The increasingly diverse operations have not been accompanied by any serious reassessment of the UN's ability to manage them effectively and its reputation has been damaged, perhaps irredeemably, in Bosnia.

The Two Yemens at War

In 1990 North and South Yemen came together in a union which the Arab world and, indeed, the larger world hoped would do away with the hostility which had long existed between the two. The terms north and south in reference to Yemen are confusing to most people beyond the region because in geographical terms the two countries are side by side in an east - west relationship. The country whose capital is San'a is west (North Yemen) and that whose capital is Aden is east (South Yemen). The political distinction is that Aden itself is south of San'a.

Together, the two countries at the time of union had a population of 14,000,000. A successful coalition seemed unlikely to foreign observers if only because they had different social and political backgrounds. All they had in common was that both were dominated for three centuries by the Ottoman Turks. Aden was seized by the British in 1839, who turned it into a coaling port when the great fleets of the world were fired by coal. Also, it was a strategic base for the British when Great Britain had a vast empire.

After achieving independence from Britain in a war that ended in 1967, the South became the first Arab Marxist state. The North had thrown off the Turks after World War I, conservative tribes gained power and ruled it ever after. Following the collapse of the Soviet Union, South Yemen's sponsor, in 1990 a merger came about. The two countries then had more in common in economic terms - oil revenues and the money both of them made from the wages Yemeni expatriates earned in other Arab states' oilfields, notably Iraq.

The leader of the North, President Ali Abdullah Saleh, and the leader of the South, Vice President Ali Salem al-Beidh, bickered incessantly. Neither trusted the other and they refused to merge their economies and their armies. Nevertheless, in 1993 Yemen's voters elected their first parliament, a remarkable exercise in democracy for a region infamous for its despotic authoritarianism. Saleh spoilt the experiment by awarding 21 of the 31 cabinet seats to his own party and to a fundamentalist leader from the North. Soon after this the fundamentalist demanded the repeal of socialist-sponsored legislation. Angry, Al-Beidh left San'a for Aden, swearing never to return.

His departure effectively crippled the government because bills could not be passed and a budget could not be presented. Law and order broke down and tourists, oil executives and diplomats were kidnapped. Inflation exceeded 100 per cent in March 1994.

In February Jordan had attempted to reconcile the two factions but each leader was making accusations against the other. Al-Beidh accused Saleh of stealing oil revenues from a new field in a southern province. Saleh accused Al-Beidh of accepting and hiding vast sums of money from King Fahd of Saudi Arabia, Yemen's neighbour. The gifts seemed unlikely since Fahd remains angry with Yemenis following the Gulf War. Saleh's support of Saddam Hussein in the war so infuriated Fahd that he threw out of Saudi Arabia about one million Yemeni workers.

In April 1994, a North Yemen tank brigade attacked and defeated a force from the South in a region north-west of San'a. This was the outbreak of war, which cannot truly be called civil war since the two countries have always been separate. Most of the fighting for the first few weeks was in the wild mountains through which passes the border between North and South. Both claimed that they would soon achieve victory but previous conflicts in the region have lasted for years.

On 14 May a Scud missile hit a residential area of San'a, though it had obviously been aimed at the nearby presidential palace. Twenty-five people were killed, the first known civilian casualties in the war. While hostilities continued in the sands and the hills both countries sent emissaries to Arab countries to seek support. Neither side asked for mediation by some Arab leader.

Remarkably, most Yemenis regard themselves as one nation, according to foreign diplomats, but the all-powerful leaders, Al-Beidh and Saleh, and their cliques hate each other with an animosity that is hard to believe.

Militarily both sides are fairly even. The South has vast supplies of material left behind by the former Soviet Union, while the North has always been well supplied by Saudi Arabia. The war is yet another example of the instability inherent in the Arab world.

References

1. A senior defence analyst at the Tel Aviv University centre for Strategic Studies, Dr. Danny Leshem, said: 'Algeria is at an advanced stage in terms of its ability to produce plutonium and it poses a considerable threat. The greater the proliferation of nuclear research and development sites throughout the Muslim world, the less Israel's ability to put an end to this growing threat.' Leshem was obliquely referring to the Israeli air strike which destroyed the Osirak reactor in Iraq, in 1981. He was speaking to a foreign journalist.
2. US Congressman Stephen Solarz had talks with North Korean leaders, including President Kim Il-sung, in December 1991. Chairman of the House sub-committee on Asian and Pacific Affairs, Solarz said: 'If the nuclear issue is not resolved it could lead to grave consequences. The clock is ticking. Time is running out.' Press conference, Washington, 23 December 1991.